Chitosan and Alginate Frontiers - Innovations and Applications in Life Sciences, Food, and Industry

Edited by Lacramioara Popa, Md. Abul Kalam Azad, Mihaela Violeta Ghica, Cristina Elena Dinu-Pîrvu, Ariana Hudita and Mireia Mallandrich Miret

Published in London, United Kingdom

Chitosan and Alginate Frontiers - Innovations and Applications in Life Sciences, Food, and Industry
http://dx.doi.org/10.5772/intechopen.1006245
Edited by Lacramioara Popa, Md. Abul Kalam Azad, Mihaela Violeta Ghica, Cristina Elena Dinu-Pîrvu, Ariana Hudita and Mireia Mallandrich Miret

Contributors
Cristina-Elena Dinu-Pîrvu, Dilyana Zvezdova, Hannis Fadzillah Mohsin, Ikram Ullah Khan, Irina-Alexandra Dumitrescu, Isha Murtaza, Kamalesh Balakumar Venkatesan, Khuriah Abdul Hamid, Ku Marsilla Ku Ishak, Lăcrămioara Popa, Manoj Kumar Srinivasan, Mashani Mohamad, Mihaela Violeta Ghica, Muhammad Irfan, Muhammad Yar, Nivedha Jayaseelan, Nurfarahin Zainal, Sajid Asghar, Saravanan Alamelu, Sohail Anjum Shahzad, Syed Haroon Khalid, Tanzeela Masood, Wan Azman Wan Ismail, Yazmin Bustami, Zulfiqar Ali Khan

First published in London, United Kingdom, 2026 by IntechOpen
IntechOpen is the global imprint of INTECHOPEN LIMITED, registered in England and Wales, registration number: 11086078, 167-169 Great Portland Street, London, W1W 5PF, United Kingdom

For EU product safety concerns: IN TECH d.o.o., Prolaz Marije Krucifikse Kozulić 3, 51000 Rijeka, Croatia, info@intechopen.com or visit our website at intechopen.com.

British Library Cataloguing-in-Publication Data
A catalogue record for this book is available from the British Library

Chitosan and Alginate Frontiers - Innovations and Applications in Life Sciences, Food, and Industry
Edited by Lacramioara Popa, Md. Abul Kalam Azad, Mihaela Violeta Ghica, Cristina Elena Dinu-Pîrvu, Ariana Hudita and Mireia Mallandrich Miret
p. cm.
Print ISBN 978-0-85466-263-0
Online ISBN 978-0-85466-262-3
eBook (PDF) ISBN 978-0-85466-264-7

Meet the editors

Lăcrămioara Popa is a Professor, Ph.D., Dr. Habil. in Pharmacy, Department of Physical and Colloidal Chemistry, Faculty of Pharmacy, Carol Davila University of Medicine and Pharmacy, Bucharest, Romania. She is a researcher at the Innovative Therapeutic Structures Research and Development Center and the President of the Ethics and Quality Assurance Commission at the Faculty of Pharmacy at the same university. She has 32 years of expertise in integrating the Quality by Design strategy into the development of drug delivery systems, surface characterization of raw materials and pharmaceutical systems, and biopolymer chitosan for medical applications. She has published over 75 ISI articles and has an H-index of 21.

Mihaela Violeta Ghica is Professor, Ph.D., Dr. Habil. in Pharmacy at the Department of Physical and Colloidal Chemistry, Faculty of Pharmacy, Carol Davila University of Medicine and Pharmacy, Bucharest, Romania. She is a researcher at the Innovative Therapeutic Structures Research and Development Center at the same university. She holds a Master's degree in Physical Chemistry and Applied Radiochemistry from the Faculty of Chemistry at the University of Bucharest. She has over 23 years of expertise in developing biomaterials for precision treatment, advanced topical formulations, statistical experimental design for drug delivery systems, and biotechnological process optimization. She has published over 100 ISI articles and has an h-index of 25.

Cristina-Elena Dinu-Pîrvu is a Professor, Ph.D., and Dr. Habil. in Pharmacy, currently affiliated with the Department of Physical and Colloidal Chemistry, Faculty of Pharmacy, at the Carol Davila University of Medicine and Pharmacy in Bucharest, Romania. At the same institution, she holds two directorship roles: Director of the Innovative Therapeutic Structures Research and Development Center and Director of the Doctoral School. She has 32 years of expertise, specifically in the design and development of bioactive compound delivery systems for personalized medicine, functionalized nanoparticulate systems for theragnostic applications, and the valorization of the biopharmacological potential of medicinal and aromatic plants. Her publication record includes more than 100 ISI articles, contributing to an H-index of 27.

Dr. Md. Abul Kalam Azad (A. K. Azad) is a Senior Lecturer at the Department of Pharmaceutical Technology, University College, MAIWP International, Malaysia, and an expert in polymers, biomaterials, Electrohydrodynamic (EHD) encapsulation technologies, and Electrospray for micro/nano particles synthesis for targeted drug delivery. Recognized in Stanford University's Top 2% Scientists (2022-present). Dr. Azad serves as Guest Editor (Elsevier, MDPI) and Editor-in-Chief of Research and Review: Drugs and Drug Development. He has reviewed 500+ articles across major publishers and delivered keynote talks worldwide. He is also a recipient of multiple innovation awards and holds a US EB-1A Green Card for extraordinary scientific contributions. He has authored 100+ publications, 20 book chapters, and several patents, with an h-index of 27.

Dr. Mireia Mallandrich Miret is a Tenure-track lecturer at the University of Barcelona. She has been involved in research since 2008 in the field of developing and evaluating topical delivery systems, while still an undergraduate student. She gained her expertise in the field through her Doctoral studies, Postdoctoral positions, Adjunct Professorship, work in the pharmaceutical industry, and her current position as a Tenure-Track Lecturer. Dr. Mallandrich is a committed professor who supervises Degree Students' Final Projects, Master's Students' Final Projects, and Doctoral Theses.

Ariana Hudita, Ph.D., serves as a Scientific Researcher within the Department of Biochemistry and Molecular Biology at the University of Bucharest. With a deep commitment to cancer research, she dedicates her expertise to advancing innovative strategies to enhance cancer patient care, with her research centered on pioneering methods to elevate the standards of cancer treatment. Her research interests encompass nanomedicine, in vitro 3D cancer modelling, tumor microenvironment characterization and manipulation, and the development of liquid biopsy protocols.

Contents

Preface

Chitosan, a natural biopolymer, is derived from chitin, one of the most abundant natural resources. In recent years, it has moved from being explored as a straightforward biodegradable material to becoming an important foundation for innovation across medicine, biotechnology, food, and industry. Its primary attributes, biocompatibility, antimicrobial activity, biodegradability, and simplicity of chemical modification, make it applicable in a variety of sectors, bridging the gap between environmental and industrial technologies and the life sciences.

This book, *Chitosan and Alginate Frontiers – Innovations and Applications in Life Sciences, Food, and Industry*, brings together recent advancements across multiple domains. It includes nanoparticles for drug delivery, materials for wound healing, and chitosan-pectin composites that are intended for tissue engineering and drug release. It also addresses smart hydrogels that are responsive to temperature and other stimuli, as well as cubosome nanocarriers that can serve as diagnostic instruments and deliver medications. These chapters collectively demonstrate how chitosan facilitates the development of contemporary, biocompatible biomaterials for medical applications.

Another important theme explored is sustainability, with chitosan investigated as a biosorbent for wastewater treatment and evaluated at both laboratory and semi-industrial scales. The results affirm that it can supplant more expensive or polluting sorbents, thereby reducing water contamination and supporting cleaner technologies.

Chitosan is demonstrated to be more than a one-purpose material by the book's integration of these domains. It is a sustainable and adaptable resource that impacts the environment, health, and industry. Each chapter emphasizes the practical applications, novel discoveries, and obstacles that must be overcome for future research and large-scale implementation. This volume will serve not only as a comprehensive overview of current progress but also as an invitation to future exploration for researchers, clinicians, engineers, and decision-makers who see in chitosan a frontier biomaterial of the 21st century.

Complementing chitosan, alginate, a naturally occurring polysaccharide extracted mainly from brown seaweeds, has gained considerable attention in biomedical and pharmaceutical research due to its remarkable biocompatibility, biodegradability, and gel-forming properties. Structurally, alginate consists of linear copolymers of β-D-mannuronic acid (M) and α-L-guluronic acid (G) residues, which determine its viscosity, mechanical strength, and ability to form ionic gels in the presence of divalent cations such as calcium.

In pharmaceutical applications, alginate is widely utilized as a matrix-forming agent, encapsulation material, and controlled-release carrier. Its ability to form hydrogels under mild conditions makes it an ideal platform for encapsulating sensitive bioactive compounds, including proteins, peptides, and live cells. Moreover, alginate-based systems have been effectively applied in wound healing, tissue engineering, and targeted drug delivery.

Recent advances have focused on the chemical modification of alginate to improve its mechanical stability, cell adhesion properties, and responsiveness to physiological stimuli. These innovations have expanded its potential beyond traditional biomedical uses, establishing alginate as a key biomaterial in the design of next-generation therapeutic systems and regenerative scaffolds.

Lăcrămioara Popa, Mihaela Violeta Ghica and Cristina-Elena Dinu-Pîrvu
Professor PhD,
Faculty of Pharmacy,
Carol Davila University of Medicine and Pharmacy,
Bucharest, Romania

Dr. Md. Abul Kalam Azad, PhD
Faculty of Pharmacy,
University College MAIWP International,
Kuala Lumpur, Malaysia

Dr. Ariana Hudiță, PhD
University of Bucharest,
Bucharest, Romania

Dr. Mireia Mallandrich Miret, PhD
University of Barcelona,
Barcelona, Spain

Chapter 1

Biomedical Potential of Chitosan Nanoparticles: Advances in Drug Delivery and Disease Management

Manoj Kumar Srinivasan, Kamalesh Balakumar Venkatesan, Nivedha Jayaseelan and Saravanan Alamelu

Abstract

Chitosan, a biopolymer derived from the deacetylation of chitin, has gained significant attention in biomedical applications due to its exceptional biocompatibility, biodegradability, and antimicrobial properties. The advent of nanotechnology has further expanded its potential through chitosan nanoparticles (CSNPs), which offer improved solubility, bioavailability, and targeted therapeutic delivery. CSNPs possess mucoadhesive properties that enhance drug transport across biological membranes, making them effective in treating cancer, diabetes, infections, and cardiovascular diseases. Additionally, their antibacterial, antifungal, and antioxidant activities contribute to food preservation, wound healing, and the mitigation of oxidative stress-related disorders. In cancer therapy, CSNPs facilitate controlled drug release, minimizing systemic toxicity and enhancing therapeutic outcomes. Their ability to interact with cellular components also makes them ideal for gene therapy and vaccine delivery. Functionalization with ligands and antibodies has further improved their specificity and bioavailability in targeted drug delivery applications, including oral, ocular, and pulmonary routes. Despite these advantages, challenges such as large-scale production, stability, and regulatory compliance must be addressed. Future research should focus on optimizing CSNP formulations for clinical applications, ensuring their safety and efficacy. This chapter explores the vast biomedical potential of CSNPs, highlighting their ability to revolutionize modern drug delivery systems.

Keywords: chitosan, nanoparticles, drug delivery, cancer therapy, antimicrobial, biocompatibility

1. Introduction

Chitosan is a polymer and polysaccharide derived from the deacetylation of chitin, one of the most abundant naturally occurring polymers found in living organisms. It is known for its biodegradability, which allows it to break down within biological systems, minimizing toxicity and ensuring biocompatibility [1]. This unique attribute makes chitosan particularly beneficial in agriculture for enhancing crop yield, managing plant infections, preventing food spoilage, and extending food shelf life. Furthermore, its antimicrobial and immunomodulatory properties have positioned it as a valuable tool in combating human and animal infections and treating a variety of diseases, including cancer, cardiovascular diseases, diabetes mellitus, and neurological disorders. In addition, chitosan plays a crucial role in improving drug and vaccine delivery, demonstrating potential in addressing infectious diseases such as COVID-19 and Lassa virus.

The versatility of chitosan as a biomaterial laid the foundation for its transition into nanotechnology-based applications, a shift that has significantly expanded its potential in medicine and pharmaceuticals. With the rise of modern nanotechnology in the 1980s [2], researchers explored ways to enhance the physicochemical properties of bioactive compounds, leading to the development of chitosan nanoparticles (CSNPs). The nanoscale formulation of chitosan has proven to be a game-changer, offering enhanced diffusibility, improved stability, and superior drug-loading capacity. This advancement has bridged the gap between chitosan's traditional uses and its evolving biomedical applications, particularly in targeted drug delivery and therapeutic interventions.

Unlike conventional chitosan, nanochitosan exhibits improved permeability and bioavailability, allowing for more efficient cellular uptake and therapeutic action. CSNPs leverage chitosan's inherent biocompatibility while overcoming the limitations of synthetic polymers, which are often associated with immune-mediated tissue rejection [3]. This has made CSNPs a promising candidate for diverse biomedical applications, ranging from food packaging where they help prevent spoilage to disease prevention and treatment. They have demonstrated remarkable efficacy against infectious diseases like COVID-19 and HIV/AIDS, as well as chronic conditions such as cardiovascular diseases, cancer, and neurological disorders.

Moreover, the functionalization of CSNPs through ionic interactions or cross-linking with other polymers further amplifies their therapeutic efficacy. These modifications enhance their interaction with biomolecules such as proteins, DNA, and microorganisms, a property particularly valuable in cancer therapy. CSNPs can disrupt the metabolic activities of cancerous cells, slowing tumor progression, while their chelating ability, which is attributed to positively charged amine groups, makes them effective in disease management. This multifaceted approach has positioned CSNPs as crucial tools not only in therapeutic drug delivery but also in diagnostic applications, enabling early disease detection and improved treatment outcomes.

Beyond their therapeutic applications, nanotechnology has redefined the way bioactive substances, including both water-soluble and insoluble medicinal compounds, are utilized in medicine. Encapsulating drugs within biodegradable nanoparticles enhances their solubility, bioavailability, and retention time, leading to sustained therapeutic effects and reduced systemic toxicity [4]. The flexibility in surface functionalization further allows for targeted drug delivery, minimizing off-target effects while optimizing drug stability [5]. Compared to traditional drug delivery methods, nanoparticle-based systems offer superior precision, controlled release, and enhanced pharmacokinetic profiles, making them a cornerstone of modern pharmacology [6].

Among the various nano-biomaterials and synthetic polymers explored for pharmaceutical applications, chitosan stands out due to its exceptional biological, physical, and chemical properties. Its bioadhesive nature enables it to interact rapidly with negatively charged mucosal membranes, prolonging drug contact time and enhancing therapeutic efficacy [7, 8]. This makes CSNPs an indispensable tool in fields such as tissue engineering, gene therapy, and nanomedicine. This chapter delves into the potential of CSNPs in revolutionizing drug delivery systems, emphasizing their transformative role in modern therapeutics.

2. Properties of chitosan

Chitosan, derived from chitin, is a versatile polysaccharide with numerous biomedical applications. Chitin, the second most abundant natural polysaccharide after cellulose [9], is the primary source for chitosan production. The quality of chitosan depends on the origin of chitin, the extraction method employed, and the degree of deacetylation achieved [10].

2.1 Sources of chitosan

Chitosan and chitin are predominantly sourced from both aquatic and terrestrial organisms, as well as specific microbial species [11]. Marine organisms, such as shrimp, lobster, crab, and squid, are significant sources of chitin. These organisms produce biowaste, which can be utilized for chitosan and chitin extraction [12, 13]. However, challenges related to seasonal availability and processing constraints hinder the constant commercial production of chitosan from aquatic biological waste [13, 14].

To overcome these challenges, terrestrial species, including mushrooms, crustaceans, and insects, have emerged as alternative sources of chitosan, particularly in the form of partially deacetylated polymers [15]. Among terrestrial organisms, silkworms and honeybees have gained attention due to their waste products being suitable for large-scale chitin and chitosan production [16]. Microbial organisms such as yeasts, molds, chrysophyte algae, ciliates, and certain bacteria (e.g., streptomycetes spores and Prosthecate bacteria stalks) also contain chitin and chitosan [17].

The microbial production of chitosan offers advantages, allowing the process to be tailored for cleaner and more uniform products with desirable properties [18]. Nevertheless, practical and economic challenges limit the use of chitosan and chitin from a wide variety of micro- and macroorganisms [19].

2.2 Biomedical properties of chitosan

Chitosan exhibits a range of properties that make it an exceptional biomaterial for biomedical applications (**Figure 1**). These include:

1. *Low immunogenicity*: Chitosan does not induce significant inflammation or immune system activation, making it ideal for medical use [20, 21].

2. *Biodegradability and biocompatibility:* Chitosan is biodegradable and biocompatible, characteristics that are essential for its biomedical applications. *In vivo*, bioenzymes facilitate the depolymerization of chitosan, leading to non-toxic, non-allergenic breakdown products such as N-acetylglucosamine and glucosamine. These products do not accumulate in the body, making chitosan a safe and versatile biomaterial for various medical uses.

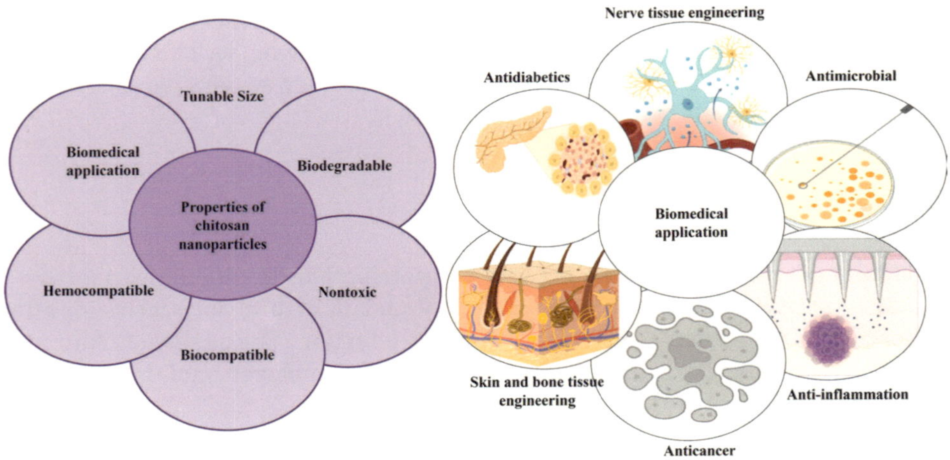

Figure 1.
Properties of chitosan nanoparticles and its biomedical applications (Own Work).

3. *Antibacterial activity*: The cationic nature of chitosan allows it to bind to bacterial surfaces, disrupting membrane permeability and inhibiting microbial growth [22]. This antibacterial effect is amplified at lower degrees of deacetylation and in acidic environments. Additionally, reducing the molecular weight of chitosan enhances its activity against Gram-negative bacteria, although it reduces its effectiveness against Gram-positive bacteria [23].

4. *Mucoadhesive properties*: Chitosan's mucoadhesive capabilities allow it to facilitate drug delivery *via* mucosal routes and enhance the adsorption of compounds that typically lack affinity for mucus [24]. It can also increase drug permeation through epithelial tissues by opening tight junctions within the epithelium [25].

5. *Wound healing and coagulation*: Chitosan has been extensively studied for its coagulation properties, promoting platelet interaction and facilitating wound healing [24]. It accelerates macrophage and neutrophil activation, granulation tissue formation, re-epithelialization, and reduces scar formation and wound contraction. As a wound dressing material, chitosan offers several beneficial properties, including chemoattraction, analgesic effects, and inherent antimicrobial activity, making it invaluable in advanced wound care [26].

2.3 Structural and chemical properties

Chitosan is a semi-crystalline polysaccharide consisting of linear chains of N-acetyl-d-glucosamine and d-glucosamine residues. Its amino group (-NH_2) imparts a cationic nature, enabling it to attract negatively charged molecules like proteoglycans, thus promoting extracellular matrix formation [27]. The hydroxyl group (-OH) enhances bonding and structural stability by interacting with positively charged molecules [28]. These functional groups allow for electrostatic interactions and versatile modifications of chitosan, improving its mechanical and physical properties [29].

Chitosan can be synthesized through the precipitation of polyanions in alkaline solutions, which imparts a wide range of medical properties, including anti-ulcer [30], wound healing [31], antibacterial activity, and cholesterol-lowering effects [32]. However, further research is required to fully understand and optimize these effects [33, 34].

2.4 Chemical modification and applications

The cationic R-NH_3^+ group in chitosan interacts with negatively charged mucosal surfaces, contributing to its mucoadhesive properties [35]. Additionally, protonated amine groups interact with cell membranes, inducing structural changes in protein-associated tight junctions, leading to transient opening [36]. The ease of chemical modification, especially at the C-2 position, allows for the development of derivatives with unique properties and applications [37].

Recent studies have shown that chitosan and its derivatives exhibit significant anticancer effects, with both *in vitro* and *in vivo* evidence supporting these findings. These effects are primarily attributed to chitosan's ability to release interleukins (IL)-1 and IL-2, enhancing the maturation and infiltration of cytolytic T-lymphocytes [38]. Furthermore, chitosan and its derivatives have demonstrated potent antioxidant activity, scavenging free radicals and protecting biomolecules from oxidative damage [39, 40]. Zhao et al. emphasized that amino and carboxyl groups in chitosan play a key role in stabilizing free radicals, thereby enhancing its antioxidant properties [41].

3. Therapeutic applications of chitosan nanoparticles

3.1 Targeted drug delivery

The recognized potential of CSNPs as drug delivery systems has created new opportunities for the development of a diverse variety of CSNPs-based delivery vehicles [42]. The United States Food and Drug Administration (FDA) recognizes chitosan's biocompatibility and classifies it as GRAS (Generally Recognized As Safe) [43]. The inclusion of amino and hydroxyl functional groups, as well as the glycosidic link, aids in drug and DNA loading into CNPs. Additionally, the solubility of CNPs in acidic aqueous solutions allows for their synthesis using sustainable chemistry practices, avoiding harmful organic solvents [44]. Another significant advantage of CNPs is their mucoadhesive property, which supports the controlled release of drugs *in vivo* [45]. CSNPs offer versatile potential applications in drug delivery, spanning various routes such as ocular, oral, parenteral, pulmonary, nasal, buccal, and vaginal administration, as well as in specialized areas like cancer therapy and tissue engineering [46]. The following sections explore selected applications of CNPs in drug delivery.

3.2 Anticancer drug delivery

Chitosan plays key roles in cancer therapy, including gene delivery, drug delivery, and vaccine adjuvants [47]. CSNPs are promising due to their biocompatibility, biodegradability, cell membrane penetration, high drug-carrying capacity, pH-responsive release, multifunctionality, and prolonged circulation [48]. Exogenous nucleic acids can be introduced into tumor cells or their surrounding environment to modify particular genes implicated in cancer etiology [49]. Nucleic acid therapies encounter several challenges, such as targeting specific tissues, overcoming charge-based repulsion at the cell membrane, and facilitating efficient endosomal escape.

These limitations hinder their therapeutic efficacy. Therefore, the development of robust nucleic acid delivery systems is crucial to overcoming these barriers and enhancing their therapeutic potential [50, 51].

Gene delivery is generally based on viral or nonviral vectors. While viral vectors have a high transfection efficiency, their mutagenic and carcinogenic potential restricts their application in cancer gene therapy [52]. Nonviral vectors, including polymer-based carriers, liposomes, and nanoparticles, have gained attention as safer, more feasible alternatives for gene transfer due to advantages like ease of fabrication, reduced immunogenicity, and better control over drug delivery [53]. Liposomes are effective gene delivery systems but suffer from low encapsulation efficiency, toxicity, short shelf life, and instability. Cationic polymers offer enhanced transfection efficiency and high encapsulation capacity and aid endosomal escape through the proton sponge effect [54, 55].

Chitosan's non-toxic nature as a gene delivery vehicle, as well as its excellent physicochemical properties that facilitate nucleic acid delivery and gene transport, have led to increased research into the potential of chitosan as a gene delivery vehicle in cancer treatment [56].

Chitosan's free amine groups enable easy functionalization with chemotherapeutic agents. For example, doxorubicin (DOX) was linked to chitosan *via* a succinic anhydride spacer, forming carboxylic acid that interacted with DOX's amine groups. This carboxylic acid was then attached to chitosan's amine groups using carbodiimide chemistry, resulting in chitosan-DOX nanoparticles that self-assembled in an aqueous solution at room temperature [57]. However, increasing the DOX amount decreased the conjugation effectiveness with chitosan. Additionally, trastuzumab, a monoclonal antibody targeting Her2+, was linked to chitosan-DOX nanoparticles through lysine residue thiolation, followed by thiol-chitosan binding. Trastuzumab-conjugated chitosan-DOX nanoparticles were more selective to Her2+ cancer cells than free medicines or chitosan-DOX nanoparticles, resulting in higher absorption. Furthermore, these conjugated nanoparticles showed a remarkable ability to distinguish between Her2+ and Her2- cancer cells, indicating their potential for active targeted drug delivery.

Another approach involved the development of pH-sensitive bovine serum albumin-stabilized graphene (BSG)/chitosan nanocomposites for efficient drug delivery in cancer treatment. These nanocomposites were created by electrostatic interactions between chitosan and BSG, and they include DOX as an anticancer drug. The characterization demonstrated that CSNPs were effectively decorated on BSG, which improved drug encapsulation efficiency. The study demonstrated that BSG reduced burst release and promoted controlled DOX release at acidic pH levels typical of the tumor microenvironment. The release profile indicated sustained drug delivery over 28 days, with optimal performance at 2 wt% BSG. Release mechanism analysis revealed uniform and sustained drug release, offering significant potential for improved cancer therapy. Furthermore, the sustained release of DOX effectively suppressed cancer cell proliferation under acidic conditions, demonstrating the efficacy of the nanocomposite in combating tumors. These findings underscore the potential of pH-sensitive nanocomposites for targeted drug delivery in cancer therapy (**Figure 2**) [58].

For example, glyceryl monooleate-chitosan core-shell nanoparticles generated by an emulsification-evaporation technique were used to encapsulate paclitaxel, a hydrophobic medication. This formulation significantly increased paclitaxel's anticancer efficacy, resulting in a 1000-fold decrease in IC50 (half-maximal

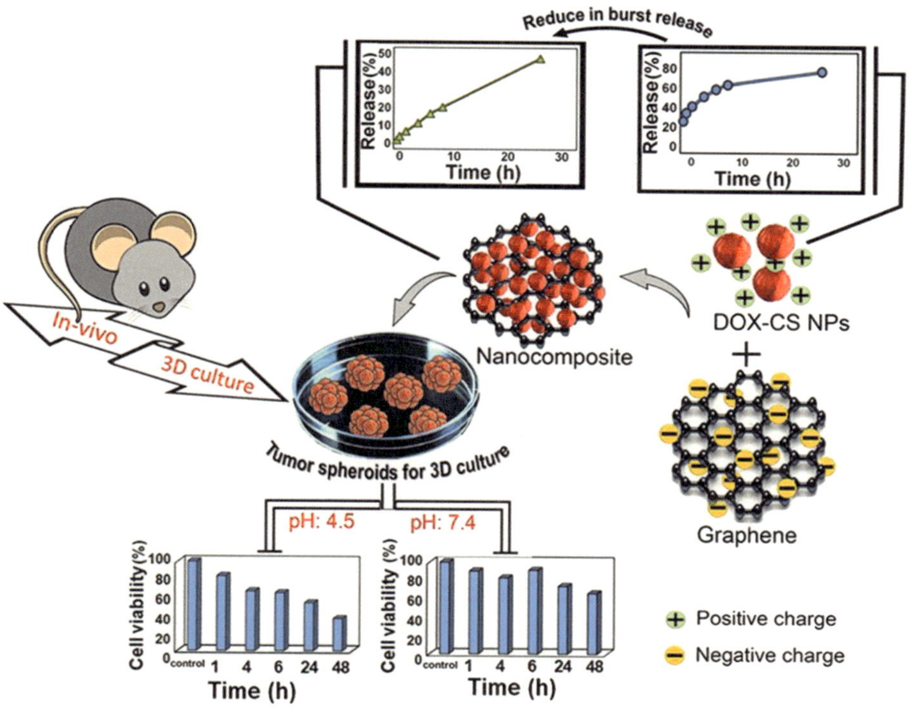

Figure 2.
Schematic illustration of the fabrication process of BSG/chitosan nanocomposites, a comparative analysis of their release profile with chitosan nanoparticles, and an investigation of their behavior in a 3D culture model [58].

inhibitory concentration). Chitosan compounds with hydrophobic drug encapsulation capabilities have also been produced to enhance the delivery of medications that are poorly water soluble [59]. Kim et al. developed amphiphilic chitosan derivatives for paclitaxel delivery by combining glycol chitosan with 5β-cholanic acid, resulting in hydrophobic glycol CSNPs (HCG). These HCG nanoparticles demonstrated an 80% drug-loading efficiency for paclitaxel, outperforming the standard Cremophor EL formulation. Additionally, the HCG nanoparticles exhibited minimal cytotoxicity [60].

Targeted drug delivery using CSNPs conjugated with tumor-specific ligands has been a promising strategy for cancer therapy. These CSNPs are designed to target overexpressed surface receptors on cancer cells, facilitating receptor-mediated endocytosis. As receptor expression levels vary with cancer types, designing customized CSNPs requires a thorough understanding of receptor profiles. Additionally, ChNPs with ligands linked *via* pH-cleavable bonds disassemble under the acidic pH of endo-lysosomes, ensuring drug release directly into the cytoplasm [61, 62].

In cancer vaccine development, adjuvants play a critical role in enhancing immune responses. Given the possible negative effects of conventional adjuvants, researchers have looked into safer alternatives such as polysaccharides produced from plants, animals, and fungus. Chitosan is regarded as an excellent vaccine adjuvant due to its cationic nature, biocompatibility, safety, and capacity to act as an antigen carrier [63, 64]. While its immunostimulatory qualities have been recognized for more than two decades, its promise as a non-toxic and effective adjuvant in cancer vaccines has just lately received widespread attention [64].

Chitosan's bioadhesive properties enhance cellular uptake, leading to robust systemic and mucosal immune responses. It stimulates both cell-mediated and humoral immunity, demonstrating immunological efficacy comparable to incomplete Freund's adjuvant and superior to aluminum hydroxide (Imject™ Alum), a widely used immunoadjuvant [65, 66]. Moreover, chitosan prolongs antigen retention at the administration site, enhancing immune response efficiency. Zaharoff et al. demonstrated that over 60% of antigens remained at the subcutaneous injection site even after 7 days, reducing the need for multiple vaccine doses [67].

CSNPs enhance OVA-induced Th1 and Th2 immune responses in mice by increasing cytokine levels (IL-2, IFN-γ, and IL-10) and stimulating natural killer cell activity. These results suggest that chitosan could be a safe and effective immunoadjuvant for cancer vaccines, promoting both cellular and humoral immune responses [68].

Sultan et al. [69] developed cisplatin-loaded CCNPs and cisplatin-loaded CSNPs surface-linked to rituximab (mAbCSNP) as innovative drug delivery systems. The formulations exhibited optimal physicochemical properties, including favorable zeta potential, particle size, and polydispersity index. Characterization through DSC and XRD confirmed their structural stability. *In vitro* drug release profiles demonstrated sustained and controlled release, following zero-order kinetics. CCNP showed significant cytotoxicity against MCF-7 human breast cancer cells, with a low IC50 value, inducing apoptosis and necrosis in a dose-dependent manner. Despite its advanced design, mAbCSNP lacked efficacy due to rituximab's specificity limitations for MCF-7 cells. These findings underscore the potential of CSNP as a promising delivery system for cisplatin, providing insights into nanocarrier-based therapeutic strategies for cancer treatment (**Figure 3**).

3.3 Vaccine delivery

Chitosan, a biodegradable biopolymer, has demonstrated significant potential in enhancing peptide adsorption and transport across the nasal epithelium, owing to its

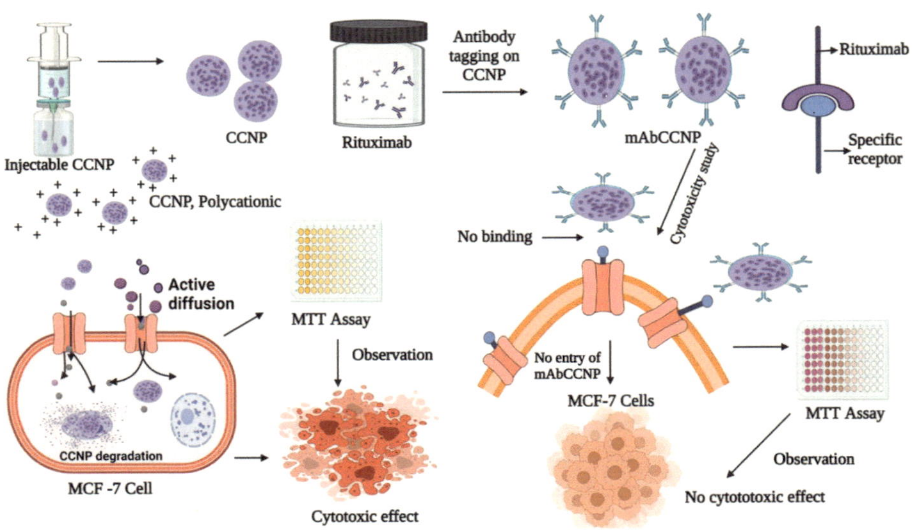

Figure 3.
Schematic representation of the CCNP and their mechanistic approach on breast cancer cells. CCNP: Cisplatin-loaded CSNPs; mAbCSNP: Rituximab conjugated cisplatin-loaded CSNPs [69].

unique mucoadhesive and osmotic properties. These properties enable it to interact with nasal tissue, facilitating the transport of macromolecules across the mucosal barrier [41]. For example, chitosan microspheres administered nasally in mice significantly improved both local and systemic immune responses to diphtheria toxoids. This highlights the relevance of the mucoadhesive properties of CSNPs in enhancing vaccine delivery.

Recent research involving both *in vitro* and *in vivo* models has further demonstrated the effectiveness of glucomannan-modified CSNPs in oral administration. Lyophilized CSNPs not only preserved the biological activity of mediators but also inhibited antigen growth. Moreover, glucuronidation of CSNPs significantly enhanced both mucosal (secretory IgA) and systemic immune responses, including serum IgG levels and cell-mediated responses (IFN-γ and IL-2). The direct connection between the mucoadhesive properties of CSNPs and their ability to boost immune responses is evident in these findings.

In another study, chitosan-dextran sulfate nanoparticles containing pertussis toxin were administered intranasally to mice, which enhanced the absorption of IgA-based nanoparticles through nasal membranes or M cells. This further supports the notion that CSNPs, with their mucoadhesive properties, facilitate the efficient uptake of antigens. Additionally, in poultry, nanoparticles containing Newcastle disease virus/influenza B virus (NDV/IBV) improved both IgG and IgA responses and enhanced lymphocyte proliferation after intranasal inoculation [70].

Chitosan has also been explored in intranasal DNA vaccination, where it was found to stimulate IFN-producing T-cells in the lungs. Furthermore, DNA carriers incorporating polyethyleneimine have been shown to enhance gene delivery by 1000-fold across the respiratory tract [71]. These studies underscore the diverse therapeutic applications of CSNPs, particularly in mucosal vaccine delivery, where their mucoadhesive and osmotic properties play a crucial role in improving immune responses.

3.4 Gene therapy delivery

Gene therapy, which includes the cellular delivery of nucleic acids, has experienced tremendous developments due to its considerable potential to cure genetic problems by introducing unique genes (RNA and DNA) into target cells for transgenic expression [72]. Effective gene therapy requires a delivery platform that is safe, efficient, and capable of overcoming various obstacles such as immunological barriers, extracellular enzymatic digestion, and cellular challenges, which are pertinent to both viral and nonviral delivery systems. Chitosan and its derivatives are commonly used as nonviral gene delivery vectors due to their low immunogenicity, mucosal adhesion, nontoxicity, availability, ease of production and administration, and capacity to carry large DNA without size limitations [73]. Numerous studies have explored chitosan-based gene vectors [72]. Among viral vectors, adeno-associated virus (AAV) is commonly used, though it rarely integrates into the host genome, despite occasional aggregation [74]. Nonviral vectors offer advantages for gene delivery because they lack genotoxicity, are scalable, and exhibit structural flexibility [75]. Chitosan-based gene vectors, along with their derivatives, continue to be a focal point of research and development in gene delivery [76]. Piri-Gharaghie et al. [77] study highlights the significance of CSNPs as efficient gene delivery vehicles due to their ability to form stable complexes with plasmid DNA (pDNA). The optimized pDNA/CSNP ratio (1:2 w.w) ensures enhanced gene uptake, as demonstrated in BALB/c mice. Advanced characterization techniques confirmed the structural integrity, stability, and plasmid localization within the nanocomplex. The findings provide valuable insights into improving targeted gene therapy applications. Khan et al. [78] study holds significant importance in the field of brain tumor therapy by addressing the critical challenge of crossing the blood-brain barrier (BBB) for effective gene delivery. The development of biocompatible CNPs as gene carriers presents a safer and more efficient alternative to conventional treatments. The nanoparticles demonstrated superior transfection efficiency and lower cytotoxicity *in vitro*, highlighting their potential for targeted therapy. *In vivo* studies further confirmed the ability of CNPs to penetrate the BBB, making them promising candidates for brain disease therapeutics. The study's findings suggest that CNPs can enhance gene delivery with improved cellular uptake and biocompatibility. This novel approach paves the way for advanced gene therapy applications, particularly in treating brain tumors. Ultimately, the research contributes to overcoming current therapeutic limitations, offering a new avenue for precision medicine in neuro-oncology.

3.5 Anti-diabetes drug delivery

CSNPs have demonstrated high promise for medication delivery in diabetes and other disorders. Amphiphilic CSNPs containing vitamin B12 were created to increase the oral bioavailability of scutellarin in type 2 diabetes-induced retinopathy. This formulation increased permeability and cellular uptake in Caco-2 cells, resulting in greater absorption. *In vivo* experiments on Sprague-Dawley rats revealed that NPs enhanced bioavailability, which alleviated diabetes-induced structural alterations in intra-retinal neovessels and decreased angiogenesis protein production, including VEGF. Furthermore, Rong et al. found that insulin-loaded chitosan NPs/PLGA-PEG-PLGA hydrogel had neuroprotective benefits in diabetic retinopathy when injected subconjunctivally [79].

In another work, Veera et al. addressed curcumin's limited bioavailability and instability by creating a new nanohybrid scaffold for tissue regeneration that included curcumin-chitosan NPs embedded in a collagen-alginate matrix. The NPs improved curcumin solubility and stability, and the nanohybrid showed excellent biocompatibility, anti-inflammatory characteristics, and cell adhesion and proliferation—all of which are important for tissue regeneration in diabetic wound healing. Rahbarian et al. investigated insulin-loaded thiolated N-triethylchitosan NPs for buccal administration, finding 98% insulin release *in vitro* and 96% permeability through the buccal mucosa in *ex vivo* experiments, indicating that this is a promising method for buccal insulin delivery. Similarly, Omid et al. created glycyl-glycine and alanylalanyline trimethyl chitosan conjugate to increase oral insulin delivery [80].

Abdel-Moneim et al. [81] study highlights the significance of developing polydatin-loaded chitosan nanoparticles (PD-CSNPs) to enhance PD's therapeutic potential against type 2 diabetes. The novel formulation improves PD's bioavailability and ensures a prolonged release pattern, with less than 20% drug release after 12 hours. Molecular simulations confirmed the interaction between PD and CSNPs, supporting efficient encapsulation. Characterization techniques validated the successful formulation of PD-CSNPs with high entrapment efficiency (96.74 ± 0.39%) and an optimal particle size (144.25 ± 3.37 nm). Cytotoxicity studies demonstrated the formula's safety across various doses. The *in vivo* study further confirmed PD-CSNPs' superior antidiabetic activity compared to free PD. Overall, this research establishes CSNPs as promising, non-toxic nanocarriers for effective PD delivery in diabetes treatment.

Alahmer et al. [82] study highlights the significance of chitosan nanoparticles (CS NPs) as an effective delivery system for *Balanites aegyptiaca* (BA) extract in diabetes treatment. The formulation enhances BA extract's bioavailability, stability, and controlled release, maximizing its therapeutic effects. BA Ex-CS NPs demonstrated significant antidiabetic activity by increasing insulin secretion and reducing fasting blood glucose and triglyceride levels. The nanoparticles promoted pancreatic beta-cell regeneration due to their strong antioxidant and anti-inflammatory properties. GC-MS, HPLC, and ICP analyses confirmed the presence of bioactive compounds responsible for the observed therapeutic effects. The study establishes CS NPs as a promising approach to improving herbal medicine efficacy in diabetes management. Overall, BA Ex-CS NPs offer a novel and effective strategy for combating diabetes mellitus.

Wardani et al. [83] study highlights the cardioprotective role of chitosan nanoparticles in diabetic-induced cardiac cell damage through antioxidative stress and antiapoptotic mechanisms. Chitosan nanoparticles significantly reduced oxidative stress markers (MDA) and apoptotic protein expression (Caspase-3) while enhancing antioxidant enzyme levels (SOD, GPx) and Nrf2 expression. The treatment also improved cardiac cell integrity by increasing Bcl-2 expression and reducing necrosis. Characterization confirmed the nanoparticles' rough surface and optimal size for effective cellular interaction. Histological analysis demonstrated preserved cardiac tissue structure in treated diabetic rats. The findings suggest that chitosan nanoparticles effectively mitigate diabetes-induced cardiac damage. This study supports their potential as a novel cardioprotective therapy for diabetic patients.

The encapsulation of resveratrol with chitosan and zinc oxide (CS–ZnO–RS) offers a novel approach to improve the therapeutic potential of resveratrol in managing gestational diabetes mellitus (GDM). Du et al. [84] study highlights the stability, bioavailability, and targeted delivery of resveratrol through CS–ZnO–RS, minimizing side effects. The synthesized nanoparticles, characterized by advanced techniques, demonstrated uniformity and an optimal size of 38 nm for effective cellular uptake. *In vitro* studies revealed potent α-glucosidase and α-amylase inhibition, which are critical for regulating postprandial glucose levels. The sustained release of resveratrol ensures prolonged efficacy, with 95% release within 24 hours. *In vivo* results confirmed the potential of CS–ZnO–RS in significantly reducing blood glucose levels, normalizing lipid profiles, and attenuating inflammation and endoplasmic reticulum stress markers. These findings suggest that CS–ZnO–RS can be a promising candidate for developing advanced therapeutic strategies for GDM, addressing both metabolic and inflammatory complications (**Figure 4**).

3.6 Antimicrobial drug delivery

CSNPs exhibit antibacterial mechanisms by disrupting the microbial cell membrane, leading to leakage of intracellular contents. They can interact with the negatively charged bacterial cell wall, increasing permeability. Chitosan also interferes with bacterial metabolic processes by binding to proteins and nucleic acids. Additionally, their ability to generate reactive oxygen species (ROS) contributes to bacterial cell damage and death (**Figure 5**) [85].

CSNPs have a wide range of antibacterial action against different pathogenic microorganisms. Ikono et al. [86] studied the effects of CS NPs on *Streptococcus mutans*, a bacterium linked with tooth cavities. Their findings revealed that increasing concentrations of CS NPs significantly reduced the viability of *S. mutans* cells.

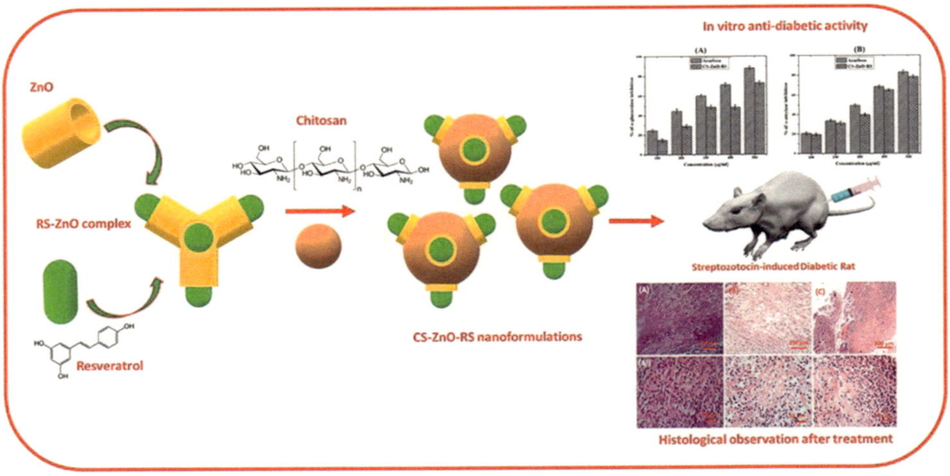

Figure 4.
Schematic representation of CS–ZnO–RS formulation, and its in vitro and in vivo antidiabetic activity [79] 85.

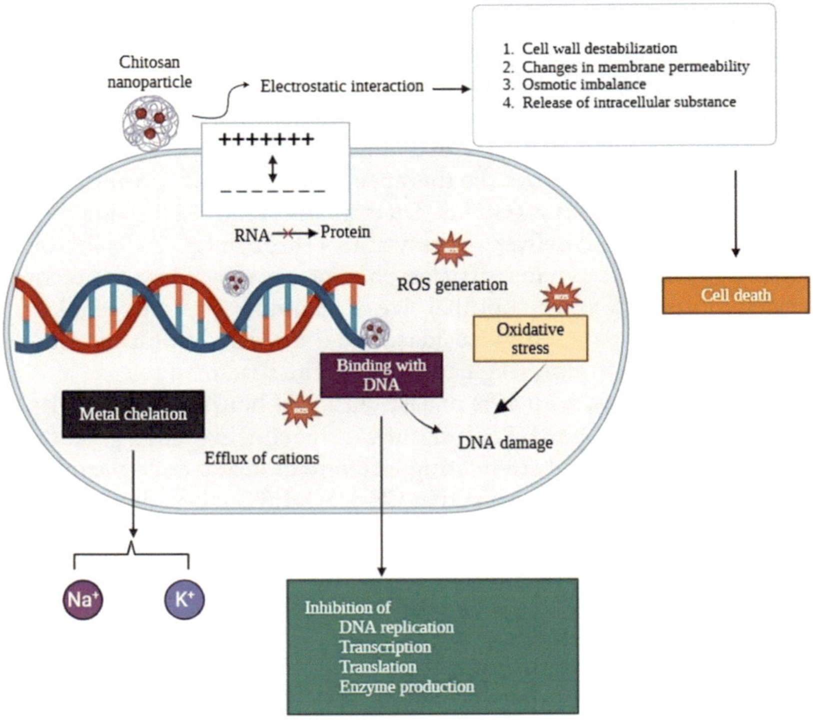

Figure 5.
Antibacterial mechanisms of chitosan nanoparticles.

In a similar study, Covarrubias et al. [87] assessed the antibacterial properties of copper-chitosan nanoparticles (Cu-CS NPs) against *S. mutans*. Cu-CS NPs demonstrated higher minimum inhibitory concentration (MIC) and minimum bactericidal concentration (MBC) values compared to copper nanoparticles (Cu-NPs) [88, 89].

Notably, Cu-CS NPs disrupted *S. mutans* adhesion and biofilm formation, which could make them promising candidates for dental plaque treatment [90, 91]. Cu-CS NPs also exhibited enhanced capacity to inhibit *S. mutans* growth on human tooth surfaces and effectively disrupted dental biofilm, showing potential for use as bactericidal agents in oral health care [87].

Hipalaswins et al. [92] investigated the antibacterial properties of CSNPs against pathogenic bacteria, including *Pseudomonas fluorescens*, *Klebsiella pneumoniae*, *Escherichia coli*, *Proteus mirabilis*, *Staphylococcus aureus*, and *Enterobacter aerogenes*. Among these, *E. aerogenes* was most susceptible to CSNPs, followed by *K. pneumoniae*, *P. fluorescens*, *E. coli*, and *P. mirabilis*, with lower toxicity observed against *S. aureus*. The results indicated that CS NPs were more effective against Gram-negative bacteria than Gram-positive ones. Nguyen et al. [93] examined the combination of CSNPs with amoxicillin against antibiotic-resistant *Streptococcus pneumoniae*, finding that the combination significantly enhanced antibacterial activity, reducing bacterial growth threefold and completely inhibiting bacterial proliferation.

Incorporating lime essential oil into CSNPs further enhanced their antibacterial properties, especially against foodborne pathogens such as *S. aureus*, *Listeria monocytogenes*, *Shigella dysenteriae*, and *E. coli*. Of these, *S. dysenteriae* showed the highest susceptibility to the CSNPs infused with lime essential oil [94]. Furthermore, CSNPs have been studied as nanocarriers for the treatment of bovine mastitis caused by *S. aureus* and coagulase-negative staphylococci [95, 96]. CSNPs displayed substantial antibacterial activity against pathogens such as *S. aureus*, *K. pneumoniae*, *E. coli*, and *Pseudomonas aeruginosa*, outperforming chitin and CS alone [97].

Tamara et al. [98] demonstrated that hybridizing chitosan with protamine enhanced the antibacterial efficacy of chitosan nanoparticles (CSNPs) against *E. coli*, but significantly reduced their effectiveness against *Bacillus cereus*. Curcumin-loaded CSNPs also showed potential in drug delivery, enhancing antibacterial responses and successfully inhibiting *S. aureus* and *P. aeruginosa* infections in mice [99]. Tran et al. [100] produced silver nanoparticles (Ag-NPs) derived from chitosan and investigated their antibacterial activities against both Gram-negative bacteria (*E. coli*, *Lactobacillus fermentum*, and *P. aeruginosa*) and Gram-positive bacteria (*S. aureus* and *Bacillus subtilis*).

Antimicrobial peptides (AMPs) like temporin B (TB) have also been employed in conjunction with CSNPs to improve antibacterial activity. Studies on TB-CSNPs exhibited bactericidal activity against clinical isolates of *Staphylococcus epidermidis*. Within 24 hours of incubation, TB-CSNPs decreased bacterial viability by altering the bacterial surface [101]. Pilon et al. [102] discovered that CSNPs, when applied as a coating, suppressed the development of mesophilic bacteria more efficiently than conventional coatings. CSNPs with a particle size of 110 nm demonstrated higher microbial suppression than conventional coatings with a particle size of 300 nm. Their findings suggest that CSNPs can be used as edible coatings to inhibit bacterial growth on fresh vegetables and fruits. The study highlighted that smaller nanoparticles exhibit increased mobility and enhanced surface interaction, which boosts their antibacterial activity against pathogens.

In studies, chitosan alone does not have significant antifungal efficacy. However, when coupled with other molecules to form nanoparticles (NP), especially with metals, its antifungal resistance against specific yeasts and molds is greatly increased [103]. This, however, is not universally agreed upon, as varying reports exist regarding the efficacy of chitosan as a standalone antifungal agent versus its nanoparticle derivative. Some researchers argue that the antifungal activity of chitosan depends on factors such as molecular mass, particle size, and the extent of acetyl removal during synthesis [104]. According to these studies, these factors enable chitosan to reduce fungal growth at specific concentrations, particularly *in vitro*. On the other hand, other studies claim that CSNPs exhibit stronger antifungal properties than chitosan alone, with these findings aligning with particle surface area and charge characteristics. The outcomes of these studies provide compelling evidence supporting the enhanced efficacy of chitosan NPs. For example, the elimination rate of fungi achieved by CSNPs was significantly higher compared to chitosan alone, showing an average 50% improvement.

Moreover, it is not just molds that are affected; yeasts are also susceptible to CSNPs. Various fungal species have been tested with CSNPs, with Candida, Fusarium, and Aspergillus species being particularly vulnerable [105, 106]. A recent survey further indicated that the fungi most susceptible to CSNPs are predominantly human pathogens [107] and plant pathogens [108]. This highlights the critical importance of CSNPs in both agricultural and health sectors. Addressing these two challenges with nanotechnology could help alleviate food scarcity while reducing mortality rates associated with invasive fungemia.

CSNPs antifungal activity is normally assessed by the extent to which fungal components under research are reduced or eliminated entirely. CSNPs act as antifungal agents by targeting a variety of fungal components, including cell membranes, spore production, mycelium growth, and cell walls [104]. One important thing to consider is particle size. Larger CSNPs have better antifungal effects, and the size correlates favorably with enhanced activity [105].

Xiaowei et al. [109] study demonstrates an eco-friendly method for creating antibacterial coatings on mobile phone glass protectors using chitosan-silver nanoparticles (ChAgNPs). The optimal chitosan concentration of 0.8% w/v resulted in AgNPs with an average diameter of 13.04 nm. Characterization techniques, including TEM, UV-vis, and FTIR spectroscopy, confirmed the successful synthesis of ChAgNPs. The optimal formulation showed a high zeta potential (+56.07 mV) and a particle size of 182.37 nm, ensuring good stability. The ChAgNP-coated glass protectors exhibited significant antibacterial activity against *E. coli*, with a reduction of 49.80% after 24 hours, though this decreased to 32.60% after 48 h (**Figure 6**).

3.7 Miscellaneous

Chitosan nanocarriers have been increasingly reported for drug delivery in various diseases, including conditions not previously discussed, such as hypertension. For instance, Saari et al. encapsulated biopeptides in CSNPs to enhance the efficiency of orally delivered antihypertensive ACE-inhibitory biopeptides. This nanoencapsulation method was created to avoid enzymatic breakdown at gastrointestinal barriers. *In vitro* studies on spontaneously hypertensive rats revealed regulated biopeptide release, with a dose-dependent reduction in blood pressure, demonstrating both gastrointestinal stability and significant antihypertensive activity of the NPs [110]. Similarly, amiodarone, an antiarrhythmic drug, was encapsulated in CSNPs with β-cyclodextrin, resulting in prolonged drug release and improved therapeutic efficacy.

Wu et al. developed folate-functionalized nanoparticles to address the low effectiveness of unencapsulated medicines in the treatment of rheumatoid arthritis. *In vitro* experiments showed that methotrexate encapsulated in folate-functionalized NPs (FA-GC-SA NPs) was more effectively absorbed by cells *via* folate

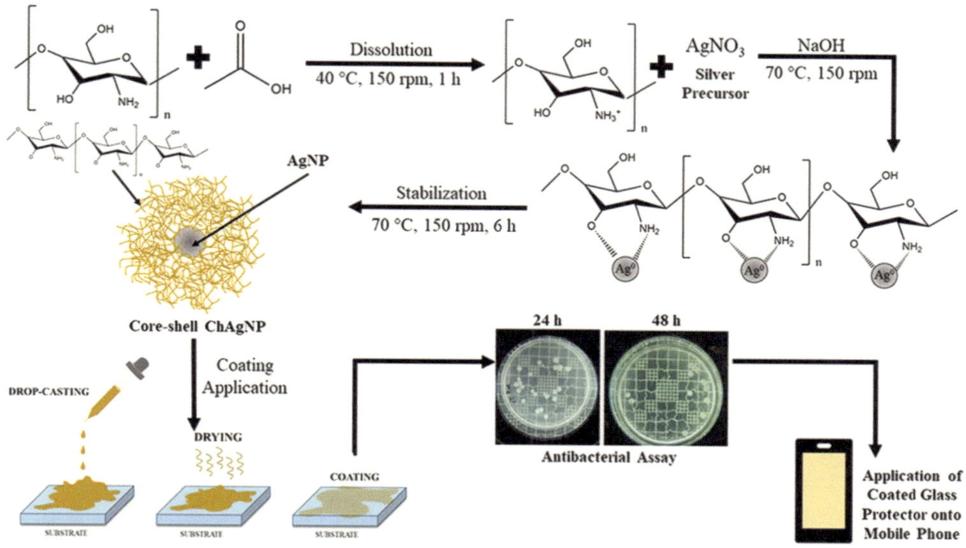

Figure 6.
Synthesis of ChAgNPs and its application of antibacterial activity on mobile phone glass protectors [109].

receptor-mediated endocytosis. *In vivo* investigations revealed a greater decrease in proinflammatory cytokines. These results revealed that the folate-functionalized CSNPs greatly improved therapeutic effectiveness. Encapsulating eugenol, an antioxidant, in CSNPs, resulted in a reduced in the expression of TGF-β and MCP-1 genes, indicating the potential for favorable therapeutic benefits in rheumatoid arthritis therapy [111].

Petkar et al. devised a unique method for administering rifampicin-loaded octanoyl CSNPs *via* the pulmonary route for TB therapy. Their investigation found that the formulation allowed for regulated drug release and had good aerosolization capabilities, which improved treatment effects. Similarly, bedaquiline encapsulated in CSNPs was tested for efficacy in both oral administration and dry powder inhalation. When compared to traditional administration techniques, the nanoformulation revealed a superior safety profile, greater therapeutic effectiveness, and enhanced target selectivity [112].

Hill et al. studied the formulation of tobramycin-loaded alginate/CSNPs as a method of minimizing noncompliance in cystic fibrosis therapy. Because of chitosan's mucoadhesive characteristics, these NPs enabled longer drug release and better effect on thick mucus in the pulmonary environment. Furthermore, functionalizing the NPs with secretory leukocyte protease inhibitor (SLPI) lowered the inflammatory response in lung infections, indicating the method's medicinal potential.

Venkatesan et al. highlighted the green synthesis of CNPs using *Plectranthus vettiveroides* root extract. The synthesized CNPs exhibited significant bioactivity, including promising antioxidant, antibacterial, and anticancer properties. Characterization techniques such as UV/vis, FTIR, and SEM confirmed the nanoparticles' spherical shape and functional groups, which are favorable for biomedical applications.
The nanoparticles demonstrated potent antibacterial activity, particularly against *Klebsiella pneumoniae*, and showed cytotoxic effects against KB cancer cells. These results suggest that *P. vettiveroides* CNPs hold potential as effective therapeutic agents in various medical applications (**Figure** 7) [113].

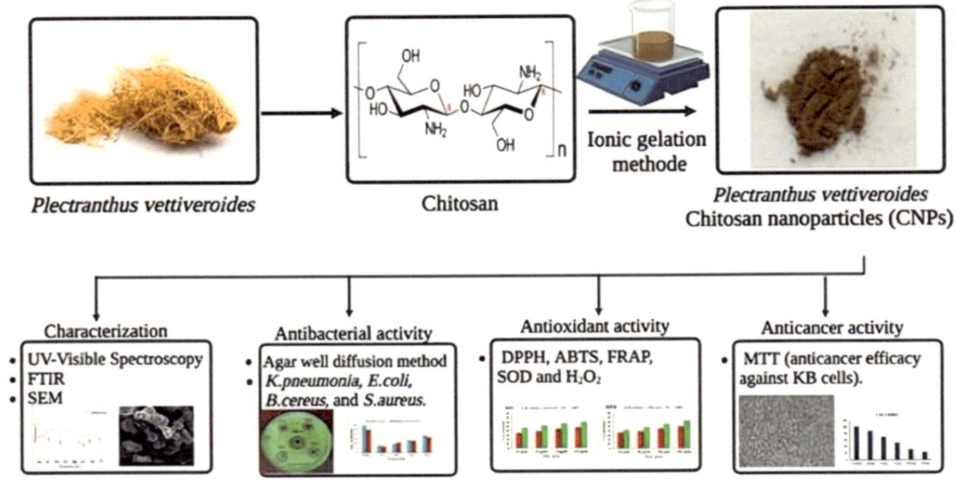

Figure 7.
Synthesis of Plectranthus vettiveroides root extract-mediated CNPs and its application on antibacterial, antioxidant, and anticancer properties [113].

4. Current clinical investigations and challenges

Chitosan, a natural polysaccharide, is widely used in various food and beverage products. The largest market for nutraceutical chitosan formulations is in Asia, particularly in Japan, Korea, and China, with the United States and Europe rapidly increasing their presence in this sector. Chitosan-based products such as NutriCology®, MicroChitosan, ChitoClear®, and Chitoseen™-F are marketed as fat-reduction and cholesterol-lowering agents for obesity and weight control. Nutritional supplements like Epakitin™ and Nutri + Gen® are designed to treat chronic renal disease in pets, particularly dogs and cats. Beyond nutraceuticals, chitosan finds diverse applications in drinks, cosmetics, agriculture, and the paper industry, with significant biological use in wound healing. Wound dressings containing chitosan include HemCon® Bandage, ChitoFlex® PRO, TraumaStat®, Svek-Patch®, ChitoGauze® PRO, Chitodine®, and Celox™. Additionally, ChitoSeat™ has emerged as a hemostatic sealant in surgical procedures to manage hemorrhages in both hard and soft tissues. **Table 1** represent the overview of biomedical application of chitosan nanoparticles.

Apart from these commercial products, there are numerous chitosan-based formulations currently undergoing clinical investigations for use in drug and vaccine delivery. However, the clinical application of chitosan faces several challenges. A major challenge with chitosan is its natural origin, which leads to variations in its structure depending on the source, extraction process, and degree of deacetylation, complicating its standardization. Furthermore, when formulated into nanoparticles (CSNPs), chitosan introduces additional complexities related to nanoparticle size, which must be carefully controlled to ensure effective long-term circulation and elimination from the body [151]. Manufacturing and scalability issues further complicate large-scale production, requiring rigorous optimization to maintain batch consistency. Regulatory hurdles also present challenges, as the approval process for CSNP-based formulations can be lengthy and complex, with potential concerns regarding their safety profile.

Type of CSNP	Formulated drug	*In Vitro* and *in vivo* tests for biological activity and drug release	Reference
Oral drug delivery of antidiabetic drugs			
CS/Dz13Scr NPs	Insulin	Insulin kinetics, *in vitro* drug release, cytotoxicity, mucus permeation, endocytic absorption study	[114]
CS NPs	Polydatin	*In vitro* drug release, cytotoxicity, antidiabetic activity in type 2 diabetic rats	[81]
Oral delivery of antihypertensive drugs			
CS NPs	Carvedilol	*In vitro* drug release, pharmacokinetics studies on rats	[115]
Oral delivery of antioxidants			
CS/Zein NPs	Resveratrol (RVT)	*In vitro* drug release, antioxidant activity, and mucoadhesion study	[116]
Oral delivery of anti-inflammatory drugs			
CS/WP-NPs	Polysaccharides from *Ophiopogon japonicus* (OJPs)	*In vitro* drug release, biocompatibility, antioxidant activity, cytotoxicity, and gene expression study.	[117]
AvrA NPs-ALG/CS MPs	Salmonella effector enzyme (AvrA)	*In vitro* drug release, reduction of inflammation in murine DSS colitis model	[118]
Oral vaccines			
β-CD/CS NPs	Ovalbumin (OVA)	*In vitro* drug release, immune response study in Balb/c mice	[119]
CS chloride NPs	Ovalbumin (OVA)	*In vitro* cell toxicity, permeability study and transepithelial electrical resistance studies.	[120]
ALG/CS NPs with Lipopolysaccharide (LPS)	HBsAg antigen	*In vitro* drug release and cytotoxicity, mucoadhesion study, and *in vivo* immunization studies	[121]
ALG/CS Coating LDHs	BSA	*In vitro* drug release, cellular uptake study.	[122]
Oral delivery of other drugs			
CS PLNs	Enoxaparin	Stability, Mucoadhesive properties and study the anticoagulant activity in rats	[123]
SA/CS and NaCAS/CS NPs	Astaxanthin (ASTX) (Hepatic fibrosis treatment)	ABTS radical scavenging assay, cytotoxicity, and anti-fibrogenic activity	[124]
Soy Lecithin/CS Hybrid NPs	Raloxifene hydrochloride (RLX)	*In vitro* drug dissolution and release, MTT assay, intestinal drug uptake, and *in vivo* pharmacokinetic studies	[125]

Type of CSNP	Formulated drug	*In Vitro* and *in vivo* tests for biological activity and drug release	Reference
Ocular drug delivery			
GCS NPs	Dexamethasone (DEX)	*In vitro* drug release, mucoadhesion, cytotoxicity, and anti-inflammatory efficacy, *in vivo* eye irritation and distribution study.	[126]
CMC/GSH/GlySar/LDHs	Dexamethasone disodium phosphate (DEXP)	*In vitro* toxicity study on human conjunctival epithelial cells, cellular uptake, precorneal retention study, and tissue distribution evaluation in rabbit eyes	[127]
TCM/LNPs	Baicalein (BAI)	*In vitro* sustained drug release, *in vivo* ocular irritation study, and pharmacokinetic evaluation	[128]
CS/Poly(ethylene glycol) Methacrylate MNPs	Bevacizumab	*In vitro* drug release kinetics, hemocompatibility, and *in vivo* antiangiogenic effect study (eye)	[129]
CS/PCL NPs	Dorzolamide (DRZ)	*In vitro* drug release, *in vivo* corneal flux, corneal hydration, *ex vivo* bioadhesion, ocular tolerance, HET-CAM test.	[130]
CS NPs	Nicotine hydrogen tartrate (NHT)	*In vitro* evaluation of nose-only inhalation device, bioactivity assessment *via* locomotor test, histopathological analysis of lung tissues	[131]
CS/PLGA NPs Coated with Chitosan	Catechin hydrate (CTH)	*In vitro* drug release, *ex vivo* permeation study on nasal mucosa, cytotoxicity, *in vivo* comparative pulmokinetic study	[132]
CS/SLNs	Rifampicin (RIF)	*In vitro* drug release, mucoadhesive properties, *in vitro* cell viability and permeability studies	[133]
Pulmonary (Inhalation) drug delivery			
Mn-TMC NPs	Etofylline (ETO)	Sustained drug release, biodegradation studies, stability, safety, aerodynamic behavior	[134]
HA/CS NPs	Ferulic acid (FA)	*In vivo* inhalation toxicity assessment	[135]
Wound healing			
CS/PEO NFs	Teicoplanin	*In vitro* drug release, antibacterial test, cytotoxicity, *in vivo* study on rat full-thickness wound model	[136]
PCL/CS NFs	Curcumin (CUR)	Antibacterial and antioxidant properties, cell viability, *in vivo* wound healing efficiency, and histological analysis	[137]

Type of CSNP	Formulated drug	*In Vitro* and *in vivo* tests for biological activity and drug release	Reference
Hydrogel Membranes with HA/PU/PVA Loaded with Cefepime-CS NPs	Cefepime	*In vitro* drug release, bacterial inhibition	[138]
CS NPs Loaded Hydrogel	*Pterocarpus marsupium* Heartwood Extract (PM)	*In vitro* drug release, antimicrobial activity, and *in vivo* wound healing in streptozotocin-induced diabetic rat models	[139]
Vaccine delivery			
CS and ALG Coated CS NPs	Hepatitis A vaccine (HAV)	HAV-specific antibody assay, lymphoproliferation assay, and effect on splenocyte proliferation in vaccinated mice	[140]
CS NPs	Aah II toxin from *Androctonus australis hector* (Scorpion Venom)	*In vitro* toxin-release study, *in vivo* immunization trial	[141]
Gene delivery			
MPC from Carbonized CTS Encapsulated ZIF-8	Luciferase-expressing plasmid (pGL3), Splice correction oligonucleotides (SCO)	Cell biocompatibility, transfection efficiency, and mechanism of uptake	[142]
LMW Mannosylated CS NPs	CpG Oligodeoxynucleotides	Cytotoxicity, cellular uptake, immunostimulatory effect, cytokine release in RAW264.7 cells, and intracellular delivery	[143]
TMC Cys, MABCMC, and CysMABC NPs	Plasmid DNA pEGFP-N1	*In vitro* DNA transfection efficiency, cytotoxicity	[144]
CS, PEI, and CMD NPs	Anti-HIV siRNA	*In vitro* cytotoxicity assay and siRNA delivery in RAW264.7 and HEK293 cells	[145]
TMC/DS or ALG NPs	hSET1 Antisense—Silencing Oligonucleotide	*In vitro* cell viability, cellular uptake, *in vivo* study on mice	[146]
CMC NPs Labeled with FITC (FITCCS/CMC)	Anti-β-Catenin siRNA	Ultrasound-triggered targeted therapy for colon cancer, *in vitro* drug release, cytotoxicity assay, cellular uptake	[147]
Guanidinylated O-CMC NPs (GOCMCS)	SiRNA Delivery	*In vitro* cell transfection studies with A549 cells, cellular uptake	[148]
CMD/TMC NPs	Co-delivery of NIK/STAT3-specific siRNA and BV6	NPs stability, *in vitro* drug release, cellular uptake, cell transfection, MTT assay, CAM assay, wound healing assay	[149]
HA/TMC NPs	IL-6- and STAT3-specific siRNAs	*In vitro* drug release, serum stability, MTT cytotoxicity, cellular uptake, transfection efficiency, colony formation, wound healing assay.	[150]

Table 1.
Represent the overview of biomedical application of chitosan nanoparticles.

In recent years, chitosan has shown promising immunostimulatory and adjuvant properties, making it a potential candidate for use as a vaccine delivery vehicle, particularly *via* the intranasal route [152, 153]. However, these immunostimulatory effects can sometimes lead to undesirable inflammatory responses, limiting its effectiveness in certain applications. Potential side effects, such as local irritation or systemic inflammation, remain a concern in clinical trials, particularly with intranasal formulations.

Despite these limitations, significant progress has been made in the clinical translation of CSNPs, with multiple formulations entering different phases of clinical trials during the last decade. Future perspectives for CSNPs are promising, as ongoing research is focused on overcoming these challenges. Recent advancements in drug delivery technologies, such as targeted nanoparticle formulations and overcoming bioavailability issues, are expected to enhance the clinical potential of chitosan. With developments in standardization procedures and overcoming these hurdles, chitosan has the potential to become the preferred delivery vehicle for a wide spectrum of clinical therapeutics.

5. Conclusion

Chitosan, a natural polysaccharide known for its exceptional biocompatibility, biodegradability, and functional adaptability, continues to demonstrate remarkable versatility in biomedical applications. Chitosan-based nanoparticles (CSNPs) have shown significant promise, particularly in antifungal therapy, targeting fungal structures such as cell membranes and spore formation, with nanoparticle size influencing efficacy. Additionally, CSNPs are effective in drug-delivery systems for conditions like hypertension, tuberculosis, rheumatoid arthritis, and cystic fibrosis, while exhibiting strong anticancer and antioxidant properties. Their eco-friendly nature also aligns with the growing demand for sustainable therapeutic approaches, seen in applications like antibacterial coatings and drug encapsulation. Despite these advantages, challenges in standardizing formulations, characterizing nanoparticles, and managing immunostimulatory effects remain. Overcoming these hurdles requires ongoing innovation in synthesis and delivery methods, along with rigorous clinical validation. The growing potential of CSNPs in tackling complex biomedical challenges marks them as a transformative tool in modern medicine. Integrating nanoparticle engineering advancements with continued research positions CSNPs for targeted, efficient, and sustainable therapies. As the field evolves, chitosan is poised to become a cornerstone material in clinical and commercial applications, driving future biomedical advancements.

Conflict of interest

The authors declare no conflict of interest.

Author details

Manoj Kumar Srinivasan[1], Kamalesh Balakumar Venkatesan[2*], Nivedha Jayaseelan[2] and Saravanan Alamelu[2]

1 Molecular Biology Lab, Department of ENT, Saveetha Medical College, Saveetha Institute of Medical and Technical Science, Chennai, Tamil Nadu, India

2 Department of Biochemistry and Biotechnology, Faculty of Science, Annamalai University, Tamil Nadu, India

*Address all correspondence to: kamalvengat2000@gmail.com

References

[1] Nagpal K, Singh SK, Mishra DN. Chitosan nanoparticles: A promising system in novel drug delivery. Chemical and Pharmaceutical Bulletin. 2010;**58**(11):1423-1430. DOI: 10.1248/cpb.58.1423

[2] Alshahrani A. The advantages of nanotechnology in medical field. International Journal of Innovative Research in Electrical, Electronics, Instrumentation and Control Engineering. 2016;**4**:1-4

[3] Jha R, Mayanovic RA. A review of the preparation, characterization, and applications of chitosan nanoparticles in nanomedicine. Nanomaterials. 2023;**13**(8):1302. DOI: 10.3390/nano13081302

[4] Ramasamy P, Dubal SV, Jeyachandran S, Pitchiah S, Kannan K, Elangovan D, et al. Control and prevention of microbially influenced corrosion using cephalopod chitosan and its derivatives: A review. International Journal of Biological Macromolecules. 2023;**242**:124924. DOI: 10.1016/j.ijbiomac.2023.124924

[5] Thanki K, Gangwal RP, Sangamwar AT, Jain S. Oral delivery of anticancer drugs: Challenges and opportunities. Journal of Controlled Release. 2013;**170**(1):15-40. DOI: 10.1016/j.jconrel.2013.04.020

[6] He M, Zhao Z, Yin L, Tang C, Yin C. Hyaluronic acid coated poly (butyl cyanoacrylate) nanoparticles as anticancer drug carriers. International Journal of Pharmaceutics. 2009;**373**(1-2):165-173. DOI: 10.1016/j.ijpharm.2009.02.012

[7] Lim C, Lee DW, Israelachvili JN, Jho Y, Hwang DS. Contact time-and pH-dependent adhesion and cohesion of low molecular weight chitosan coated surfaces. Carbohydrate Polymers. 2015;**117**:887-894. DOI: 10.1016/j.carb-pol.2014.10.033

[8] Lim C, Hwang DS, Lee DW. Intermolecular interactions of chitosan: Degree of acetylation and molecular weight. Carbohydrate Polymers. 2021;**259**:117782. DOI: 10.1016/j.carbpol.2021.117782

[9] Karthick Rajan D, Mohan K, Rajarajeswaran J, Divya D, Kumar R, Kandasamy S, et al. β-Chitin and chitosan from waste shells of edible mollusks as a functional ingredient. Food Frontiers. 2024;**5**(1):46-72

[10] Younes I, Sellimi S, Rinaudo M, Jellouli K, Nasri M. Influence of acetylation degree and molecular weight of homogeneous chitosans on antibacterial and antifungal activities. International Journal of Food Microbiology. 2014;**185**:57-63. DOI: 10.1016/j.ijfoodmicro.2014.04.029

[11] Kamerling JP, Boons GJ. Comprehensive Glycoscience: From Chemistry to Systems Biology. Oxford: Elsevier; 2007

[12] Teng WL, Khor E, Tan TK, Lim LY, Tan SC. Concurrent production of chitin from shrimp shells and fungi. Carbohydrate Research. 2001;**332**(3):305-316. DOI: 10.1016/S0008-6215(01)00084-2

[13] Nwe N, Stevens WF. Production of chitin and chitosan and their applications in the medical and biological sector. Recent Research in Biomedical Aspects of Chitin and Chitosan. 2008;**978**:161

[14] Crestini C, Kovac B, Giovannozzi-Sermanni G. Production and isolation of chitosan by submerged and solid-state fermentation from Lentinus edodes. Biotechnology and Bioengineering. 1996;**50**(2):207-210. DOI: 10.1002/bit.260500202

[15] Goy RC, Britto DD, Assis OB. A review of the antimicrobial activity of chitosan. Polímeros. 2009;**19**:241-247. DOI: 10.1590/S0104-14282009000300013

[16] Sudha PN, Gomathi T, Aisverya S. Recent research in the applications of chitin, chitosan and oligosaccharides. In: Green Polymers and Environmental Pollution Control. New York: Apple Academic Press; 2015. pp. 303-333

[17] Gooday GW. The ecology of chitin degradation. In: Advances in Microbial Ecology. Boston, MA: Springer US; 1990. pp. 387-430. DOI: 10.1007/978-1-4684-7612-5_10

[18] Santos VP, Marques NS, Maia PC, Lima MA, Franco LD, Campos-Takaki GM. Seafood waste as attractive source of chitin and chitosan production and their applications. International Journal of Molecular Sciences. 2020;**21**(12):4290. DOI: 10.3390/ijms21124290

[19] Hayes M, Carney B, Slater J, Brück W. Mining marine shellfish wastes for bioactive molecules: Chitin and chitosan – Part A: Extraction methods. Biotechnology Journal: Healthcare Nutrition Technology. 2008;**3**(7):871-877. DOI: 10.1002/biot.200700197

[20] Kong M, Chen XG, Xing K, Park HJ. Antimicrobial properties of chitosan and mode of action: A state of the art review. International Journal of Food Microbiology. 2010;**144**(1):51-63. DOI: 10.1016/j.ijfoodmicro.2010.09.012

[21] Bugnicourt L, Ladavière C. Interests of chitosan nanoparticles ionically cross-linked with tripolyphosphate for biomedical applications. Progress in Polymer Science. 2016;**60**:1-7. DOI: 10.1016/j.progpolymsci.2016.06.002

[22] Yamamoto H, Kuno Y, Sugimoto S, Takeuchi H, Kawashima Y. Surface-modified PLGA nanosphere with chitosan improved pulmonary delivery of calcitonin by mucoadhesion and opening of the intercellular tight junctions. Journal of Controlled Release. 2005;**102**(2):373-381. DOI: 10.1016/j.jconrel.2004.10.010

[23] Dhevishri S, Parameswari BD, Annapoorni H, Shankar MS, Kumar R. Antimicrobial properties of green synthesized silver and chitosan nanocomposites. Bioinformation. 2023;**19**(6):745. DOI: 10.6026/97320630019745

[24] Salamat-Miller N, Chittchang M, Johnston TP. The use of mucoadhesive polymers in buccal drug delivery. Advanced Drug Delivery Reviews. 2005;**57**(11):1666-1691. DOI: 10.1016/j.addr.2005.07.003

[25] Okamoto Y, Yano R, Miyatake K, Tomohiro I, Shigemasa Y, Minami S. Effects of chitin and chitosan on blood coagulation. Carbohydrate Polymers. 2003;**53**(3):337-342. DOI: 10.1016/S0144-8617(03)00076-6

[26] Busilacchi A, Gigante A, Mattioli-Belmonte M, Manzotti S, Muzzarelli RA. Chitosan stabilizes platelet growth factors and modulates stem cell differentiation toward tissue regeneration. Carbohydrate Polymers. 2013;**98**(1):665-676. DOI: 10.1016/j.carbpol.2013.06.044

[27] Herdiana Y, Wathoni N, Shamsuddin S, Muchtaridi M.

Drug release study of the chitosan-based nanoparticles. Heliyon. 2022;**8**(1):e08674. DOI: 10.1016/j.heliyon.2021.e08674

[28] Rahmani Del Bakhshayesh A, Annabi N, Khalilov R, Akbarzadeh A, Samiei M, Alizadeh E, et al. Recent advances on biomedical applications of scaffolds in wound healing and dermal tissue engineering. Artificial Cells, Nanomedicine, and Biotechnology. 2018;**46**(4):691-705. DOI: 10.1080/21691401.2017.1349778

[29] Khan A, Alamry KA. Recent advances of emerging green chitosan-based biomaterials with potential biomedical applications: A review. Carbohydrate Research. 2021;**506**:108368. DOI: 10.1016/j.carres.2021.108368

[30] Fini A, Orienti I. The role of chitosan in drug delivery: Current and potential applications. American Journal of Drug Delivery. 2003;**1**:43-59. DOI: 10.2165/00137696-200301010-00004

[31] Azad AK, Sermsintham N, Chandrkrachang S, Stevens WF. Chitosan membrane as a wound-healing dressing: Characterization and clinical application. Journal of Biomedical Materials Research Part B: Applied Biomaterials: An Official Journal of The Society for Biomaterials, The Japanese Society for Biomaterials, and The Australian Society for Biomaterials and the Korean Society for Biomaterials. 2004;**69**(2):216-222. DOI: 10.1002/jbm.b.30000

[32] Sugano M, Watanabe S, Kishi A, Izume M, Ohtakara A. Hypocholesterolemic action of chitosans with different viscosity in rats. Lipids. 1988;**23**:187-191. DOI: 10.1007/BF02535456

[33] Rabea EI, Badawy ME, Steurbaut W, Rogge TM, Stevens CV, Smagghe G, et al. Fungicidal effect of chitosan derivatives containing an N-alkyl group on grey mould Botrytis cinerea and rice leaf blast Pyricularia grisea. Colloids and Surfaces B: Biointerfaces. 2022;**219**:112824. DOI: 10.1016/j.colsurfb.2022.112824

[34] Sanpui P, Murugadoss A, Prasad PD, Ghosh SS, Chattopadhyay A. The antibacterial properties of a novel chitosan–Ag-nanoparticle composite. International Journal of Food Microbiology. 2008;**124**(2):142-146. DOI: 10.1016/j.ijfoodmicro.2008.03.004

[35] Kockisch S, Rees GD, Young SA, Tsibouklis J, Smart JD. Polymeric microspheres for drug delivery to the oral cavity: An in vitro evaluation of mucoadhesive potential. Journal of Pharmaceutical Sciences. 2003;**92**(8):1614-1623. DOI: 10.1002/jps.10423

[36] Sinha VR, Singla AK, Wadhawan S, Kaushik R, Kumria R, Bansal K, et al. Chitosan microspheres as a potential carrier for drugs. International Journal of Pharmaceutics. 2004;**274**(1-2):1-33. DOI: 10.1016/j.ijpharm.2003.12.026

[37] Huo M, Zhang Y, Zhou J, Zou A, Yu D, Wu Y, et al. Synthesis and characterization of low-toxic amphiphilic chitosan derivatives and their application as micelle carrier for antitumor drug. International Journal of Pharmaceutics. 2010;**394**(1-2):162-173. DOI: 10.1016/j.ijpharm.2010.05.001

[38] Tokoro A, Takewaki N, Suzuki KO, Mikami T, Suzuki S, Suzuki M. Growth-inhibitory effect of hexa-N-acetylchitohexanse and chitohexaose against meth-a solid tumor. Chemical and Pharmaceutical Bulletin. 1988;**36**(2):784-790. DOI: 10.1248/cpb.36.784

[39] Park PJ, Je JY, Kim SK. Free radical scavenging activity of chitooligosaccharides by electron spin resonance spectrometry. Journal of Agricultural and Food Chemistry. 2003;**51**(16):4624-4627. DOI: 10.1021/jf034039+

[40] Anisiei A, Oancea F, Marin L. Electrospinning of chitosan-based nanofibers: From design to prospective applications. Reviews in Chemical Engineering. 2023;**39**(1):31-70. DOI: 10.1515/revce-2021-0003

[41] Zhao D, Yu S, Sun B, Gao S, Guo S, Zhao K. Biomedical applications of chitosan and its derivative nanoparticles. Polymers. 2018;**10**(4):462. DOI: 10.3390/polym10040462

[42] Malmiri HJ, Jahanian MA, Berenjian A. Potential applications of chitosan nanoparticles as novel support in enzyme immobilization. American Journal of Biochemistry and Biotechnology. 2012;**8**(4):203-219

[43] Pangestuti R, Kim SK. Neuroprotective properties of chitosan and its derivatives. Marine Drugs. 2010;**8**(7):2117-2128. DOI: 10.3390/md8072117

[44] Garg U, Chauhan S, Nagaich U, Jain N. Current advances in chitosan nanoparticles based drug delivery and targeting. Advanced Pharmaceutical Bulletin. 2019;**9**(2):195. DOI: 10.15171/apb.2019.023

[45] Bhumkar DR, Pokharkar VB. Studies on effect of pH on cross-linking of chitosan with sodium tripolyphosphate: A technical note. AAPS PharmSciTech. 2006;**7**:E138-E143. DOI: 10.1208/pt070250

[46] Mohammed MA, Syeda JT, Wasan KM, Wasan EK. An overview of chitosan nanoparticles and its application in non-parenteral drug delivery. Pharmaceutics. 2017;**9**(4):53. DOI: 10.3390/pharmaceutics9040053

[47] Prabaharan M. Chitosan-based nanoparticles for tumor-targeted drug delivery. International Journal of Biological Macromolecules. 2015;**72**:1313-1322. DOI: 10.1016/j.ijbiomac.2014.10.052

[48] Cross D, Burmester JK. Gene therapy for cancer treatment: Past, present and future. Clinical Medicine & Research. 2006;**4**(3):218-227. DOI: 10.3121/cmr.4.3.218

[49] Wang J, Lu Z, Wientjes MG, Au JL. Delivery of siRNA therapeutics: Barriers and carriers. The AAPS Journal. 2010;**12**:492-503. DOI: 10.1208/s12248-010-9210-4

[50] Sayed N, Allawadhi P, Khurana A, Singh V, Navik U, Pasumarthi SK, et al. Gene therapy: Comprehensive overview and therapeutic applications. Life Sciences. 2022;**294**:120375

[51] Nayerossadat N, Maedeh T, Ali PA. Viral and nonviral delivery systems for gene delivery. Advanced Biomedical Research. 2012;**1**(1):27. DOI: 10.4103/2277-9175.98152

[52] Thomas M, Klibanov AM. Non-viral gene therapy: Polycation-mediated DNA delivery. Applied Microbiology and Biotechnology. 2003;**62**:27-34. DOI: 10.1007/s00253-003-1321-8

[53] Kedmi R, Ben-Arie N, Peer D. The systemic toxicity of positively charged lipid nanoparticles and the role of toll-like receptor 4 in immune activation. Biomaterials. 2010;**31**(26):6867-6875. DOI: 10.1016/j.biomaterials.2010.05.027

[54] Li S, Tian X, Fan J, Tong H, Ao Q, Wang X. Chitosans for tissue

repair and organ three-dimensional (3D) bioprinting. Micromachines. 2019;**10**(11):765. DOI: 10.3390/mi10110765

[55] Sun X, Zhang N. Cationic polymer optimization for efficient gene delivery. Mini Reviews in Medicinal Chemistry. 2010;**10**(2):108-125. DOI: 10.2174/138955710791185109

[56] Mao S, Sun W, Kissel T. Chitosan-based formulations for delivery of DNA and siRNA. Advanced Drug Delivery Reviews. 2010;**62**(1):12-27. DOI: 10.1016/j.addr.2009.08.004

[57] Yousefpour P, Atyabi F, Vasheghani-Farahani E, Movahedi AA, Dinarvand R. Targeted delivery of doxorubicin-utilizing chitosan nanoparticles surface-functionalized with anti-Her2 trastuzumab. International Journal of Nanomedicine. 2011;**6**:1977-1990. DOI: 10.2147/IJN.S21523

[58] Gooneh-Farahani S, Naghib SM, Naimi-Jamal MR, Seyfoori A. A pH-sensitive nanocarrier based on BSA-stabilized graphene-chitosan nanocomposite for sustained and prolonged release of anticancer agents. Scientific Reports. 2021;**11**(1):17404

[59] Trickler WJ, Nagvekar AA, Dash AK. A novel nanoparticle formulation for sustained paclitaxel delivery. AAPS PharmSciTech. 2008;**9**:486-493. DOI: 10.1208/s12249-008-9063-7

[60] Kim JH, Kim YS, Kim S, Park JH, Kim K, Choi K, et al. Hydrophobically modified glycol chitosan nanoparticles as carriers for paclitaxel. Journal of Controlled Release. 2006;**111**(1-2):228-234. DOI: 10.1016/j.jconrel.2005.12.013

[61] Ghaz-Jahanian MA, Abbaspour-Aghdam F, Anarjan N, Berenjian A, Jafarizadeh-Malmiri H. Application of chitosan-based nanocarriers in tumor-targeted drug delivery. Molecular Biotechnology. 2015;**57**:201-218. DOI: 10.1007/s12033-014-9816-3

[62] Xu S, Olenyuk BZ, Okamoto CT, Hamm-Alvarez SF. Targeting receptor-mediated endocytotic pathways with nanoparticles: Rationale and advances. Advanced Drug Delivery Reviews. 2013;**65**(1):121-138. DOI: 10.1016/j.addr.2012.09.041

[63] Petrovsky N, Aguilar JC. Vaccine adjuvants: Current state and future trends. Immunology and Cell Biology. 2004;**82**(5):488-496. DOI: 10.1111/j.0818-9641.2004.01272.x

[64] Petrovsky N, Cooper PD. Carbohydrate-based immune adjuvants. Expert Review of Vaccines. 2011;**10**(4):523-537. DOI: 10.1586/erv.11.30

[65] Zaharoff DA, Rogers CJ, Hance KW, Schlom J, Greiner JW. Chitosan solution enhances both humoral and cell-mediated immune responses to subcutaneous vaccination. Vaccine. 2007;**25**(11):2085-2094. DOI: 10.1016/j.vaccine.2006.11.034

[66] Powell BS, Andrianov AK, Fusco PC. Polyionic vaccine adjuvants: Another look at aluminum salts and polyelectrolytes. Clinical and Experimental Vaccine Research. 2015;**4**(1):23-45. DOI: 10.7774/cevr.2015.4.1.23

[67] Zaharoff DA, Hance KW, Rogers CJ, Schlom J, Greiner JW. Intratumoral immunotherapy of established solid tumors with chitosan/IL-12. Journal of Immunotherapy. 2010;**33**(7):697-705. DOI: 10.1097/CJI.0b013e3181eb826d

[68] Wen ZS, Xu YL, Zou XT, Xu ZR. Chitosan nanoparticles act as an adjuvant

to promote both Th1 and Th2 immune responses induced by ovalbumin in mice. Marine Drugs. 2011;**9**(6):1038-1055. DOI: 10.3390/md9061038

[69] Sultan MH, Moni SS, Madkhali OA, Bakkari MA, Alshahrani S, Alqahtani SS, et al. Characterization of cisplatin-loaded chitosan nanoparticles and rituximab-linked surfaces as target-specific injectable nano-formulations for combating cancer. Scientific Reports. 2022;**12**(1):468

[70] Zhao W, Yang Y, Song L, Kang T, Du T, Wu Y, et al. A vesicular stomatitis virus-inspired DNA nanocomplex for ovarian cancer therapy. Advanced Science. 2018;**5**(3):1700263. DOI: 10.1002/advs.201700263

[71] Torrieri-Dramard L, Lambrecht B, Ferreira HL, Van den Berg T, Klatzmann D, Bellier B. Intranasal DNA vaccination induces potent mucosal and systemic immune responses and cross-protective immunity against influenza viruses. Molecular Therapy. 2011;**19**(3):602-611. DOI: 10.1038/mt.2010.222

[72] Yalcin S, Cankaya N. Chitosan-based nanocomposites for biological applications. In: Nanoclay-Recent Advances, New Perspectives and Applications. London, UK: IntechOpen; 2022. DOI: 10.5772/intechopen.106379

[73] Liu Y, Wang FQ, Shah Z, Cheng XJ, Kong M, Feng C, et al. Nano-polyplex based on oleoyl-carboxymethy-chitosan (OCMCS) and hyaluronic acid for oral gene vaccine delivery. Colloids and Surfaces B: Biointerfaces. 2016;**145**:492-501. DOI: 10.1016/j.colsurfb.2016.05.035

[74] Yin H, Song CQ, Suresh S, Wu Q, Walsh S, Rhym LH, et al. Structure-guided chemical modification of guide RNA enables potent non-viral in vivo genome editing. Nature Biotechnology. 2017;**35**(12):1179-1187. DOI: 10.1038/nbt.4005

[75] Cortez MA, Godbey WT, Fang Y, Payne ME, Cafferty BJ, Kosakowska KA, et al. The synthesis of cyclic poly (ethylene imine) and exact linear analogues: An evaluation of gene delivery comparing polymer architectures. Journal of the American Chemical Society. 2015;**137**(20):6541-6549. DOI: 10.1021/jacs.5b00980

[76] Chuan D, Jin T, Fan R, Zhou L, Guo G. Chitosan for gene delivery: Methods for improvement and applications. Advances in Colloid and Interface Science. 2019;**268**:25-38. DOI: 10.1016/j.cis.2019.03.007

[77] Piri-Gharaghie T, Doosti A, Mirzaei SA. Fabrication and characterization of pcDNA3. 1 (+) location within chitosan/nanoparticles complexes for enhanced gene delivery. Iranian Journal of Biotechnology. 2022;**20**(3):e3110. DOI: 10.30498/ijb.2022.297534.3110

[78] Khan IN, Navaid S, Waqar W, Hussein D, Ullah N, Khan MUA, et al. Chitosan-based polymeric nanoparticles as an efficient gene delivery system to cross blood brain barrier: In vitro and in vivo evaluations. Pharmaceuticals. 2024;**17**(2):169. DOI: 10.3390/ph17020169

[79] Rong X, Ji Y, Zhu X, Yang J, Qian D, Mo X, et al. Neuroprotective effect of insulin-loaded chitosan nanoparticles/PLGA-PEG-PLGA hydrogel on diabetic retinopathy in rats. International Journal of Nanomedicine. 2019;**14**:45-55. DOI: 10.2147/IJN.S184574

[80] Omid NJ, Javan NB, Dehpour AR, Partoazar A, Tehrani MR, Dorkoosh F.

In-vitro and in-vivo cytotoxicity and efficacy evaluation of novel glycyl-glycine and alanyl-alanine conjugates of chitosan and trimethyl chitosan nano-particles as carriers for oral insulin delivery. International Journal of Pharmaceutics. 2018;**535**(1-2):293-307. DOI: 10.1016/j.ijpharm.2017.11.020

[81] Abdel-Moneim A, El-Shahawy A, Yousef AI, Abd El-Twab SM, Elden ZE, Taha M. Novel polydatin-loaded chitosan nanoparticles for safe and efficient type 2 diabetes therapy: In silico, in vitro and in vivo approaches. International Journal of Biological Macromolecules. 2020;**154**:1496-1504. DOI: 10.1016/j.ijbiomac.2019.11.031

[82] Alahmer S, El-Noss M, Farid A. Preparation of chitosan nanoparticles loaded with Balanites aegyptiaca extract for treatment of streptozotocin-induced diabetes in rats. International Journal of Biological Macromolecules. 2024;**262**:130061. DOI: 10.1016/j.ijbiomac.2024.130061

[83] Wardani G, Nugraha J, Mustafa MR, Kurnijasanti R, Sudjarwo SA. Antioxidative stress and antiapoptosis effect of chitosan nanoparticles to protect cardiac cell damage on streptozotocin-induced diabetic rat. Oxidative Medicine and Cellular Longevity. 2022;**2022**(1):3081397. DOI: 10.1155/2022/3081397

[84] Du S, Lv Y, Li N, Huang X, Liu X, Li H, et al. Biological investigations on therapeutic effect of chitosan encapsulated nano resveratrol against gestational diabetes mellitus rats induced by streptozotocin. Drug Delivery. 2020;**27**(1):953-963

[85] Chandrasekaran M, Kim KD, Chun SC. Antibacterial activity of chitosan nanoparticles: A review. PRO. 2020;**8**(9):1173

[86] Wei D, Sun W, Qian W, Ye Y, Ma X. The synthesis of chitosan-based silver nanoparticles and their antibacterial activity. Carbohydrate Research. 2009;**344**(17):2375-2382. DOI: 10.1016/j.ijpharm.2017.11.020

[87] Covarrubias C, Trepiana D, Corral C. Synthesis of hybrid copper-chitosan nanoparticles with antibacterial activity against cariogenic *Streptococcus mutans*. Dental Materials Journal. 2018;**37**(3):379-384. DOI: 10.4012/dmj.2017-195

[88] Espinosa-Cristóbal LF, Martínez-Castañón GA, Martínez-Martínez RE, Loyola-Rodríguez JP, Patino-Marin N, Reyes-Macías JF, et al. Antibacterial effect of silver nanoparticles against *Streptococcus mutans*. Materials Letters. 2009;**63**(29):2603-2606. DOI: 10.1016/j.matlet.2009.09.018

[89] Besinis A, De Peralta T, Handy RD. The antibacterial effects of silver, titanium dioxide and silica dioxide nanoparticles compared to the dental disinfectant chlorhexidine on *Streptococcus mutans* using a suite of bioassays. Nanotoxicology. 2014;**8**(1):1-6. DOI: 10.3109/17435390.2012.742935

[90] Aliasghari A, Khorasgani MR, Vaezifar S, Rahimi F, Younesi H, Khoroushi M. Evaluation of antibacterial efficiency of chitosan and chitosan nanoparticles on cariogenic streptococci: An in vitro study. Iranian Journal of Microbiology. 2016;**8**(2):93

[91] Costa EM, Silva S, Tavaria FK, Pintado MM. Study of the effects of chitosan upon *Streptococcus mutans* adherence and biofilm formation. Anaerobe. 2013;**20**:27-31. DOI: 10.1016/j.anaerobe.2013.02.002

[92] Hipalaswins WM, Balakumaran M, Jagadeeswari S. Synthesis,

characterization and antibacterial activity of chitosan nanoparticles and its impact on seed germination. Journal of Academia and Industrial Research. 2016;**5**(5):65-71

[93] Nguyen TV, Nguyen TT, Wang SL, Vo TP, Nguyen AD. Preparation of chitosan nanoparticles by TPP ionic gelation combined with spray drying, and the antibacterial activity of chitosan nanoparticles and a chitosan nanoparticle–amoxicillin complex. Research on Chemical Intermediates. 2017;**43**:3527-3537. DOI: 10.1007/s11164-016-2428-8

[94] Sotelo-Boyás ME, Correa-Pacheco ZN, Bautista-Baños S, Corona-Rangel ML. Physicochemical characterization of chitosan nanoparticles and nanocapsules incorporated with lime essential oil and their antibacterial activity against food-borne pathogens. LWT. 2017;**77**:15-20. DOI: 10.1016/j.lwt.2016.11.022

[95] Taponen S, Pyörälä S. Coagulase-negative staphylococci as cause of bovine mastitis—Not so different from *Staphylococcus aureus*? Veterinary Microbiology. 2009;**134**(1-2):29-36. DOI: 10.1016/j.vetmic.2008.09.011

[96] Cardozo VF, Lancheros CA, Narciso AM, Valereto EC, Kobayashi RK, Seabra AB, et al. Evaluation of antibacterial activity of nitric oxide-releasing polymeric particles against *Staphylococcus aureus* and *Escherichia coli* from bovine mastitis. International Journal of Pharmaceutics. 2014;**473**(1-2):20-29. DOI: 10.1016/j.ijpharm.2014.06.051

[97] Ma Z, Garrido-Maestu A, Jeong KC. Application, mode of action, and in vivo activity of chitosan and its micro-and nanoparticles as antimicrobial agents: A review. Carbohydrate Polymers. 2017;**176**:257-265. DOI: 10.1016/j.carbpol.2017.08.082

[98] Tamara FR, Lin C, Mi FL, Ho YC. Antibacterial effects of chitosan/cationic peptide nanoparticles. Nanomaterials. 2018;**8**(2):88. DOI: 10.3390/nano8020088

[99] Mofazzal JM, Al Musawi S, Pirestani M, Fasihi RM, Ahmadi K, Rajayi H, et al. Curcumin-loaded chitosan tripolyphosphate nanoparticles as a safe, natural and effective antibiotic inhibits the infection of *Staphylococcus aureus* and *Pseudomonas aeruginosa* in vivo. Iranian Journal of Biotechnology. 2014;**12**(3):e1012. DOI: 10.15171/ijb.1012

[100] Tran HV, Dai Tran L, Ba CT, Vu HD, Nguyen TN, Pham DG, et al. Synthesis, characterization, antibacterial and antiproliferative activities of monodisperse chitosan-based silver nanoparticles. Colloids and Surfaces A: Physicochemical and Engineering Aspects. 2010;**360**(1-3):32-40. DOI: 10.1016/j.colsurfa.2010.02.007

[101] Piras AM, Maisetta G, Sandreschi S, Gazzarri M, Bartoli C, Grassi L, et al. Chitosan nanoparticles loaded with the antimicrobial peptide temporin B exert a long-term antibacterial activity in vitro against clinical isolates of *Staphylococcus epidermidis*. Frontiers in Microbiology. 2015;**6**:372. DOI: 10.3389/fmicb.2015.00372

[102] Pilon L, Spricigo PC, Miranda M, de Moura MR, Assis OB, Mattoso LH, et al. Chitosan nanoparticle coatings reduce microbial growth on fresh-cut apples while not affecting quality attributes. International Journal of Food Science & Technology. 2015;**50**(2):440-448. DOI: 10.1111/ijfs.12616

[103] Al-Zahrani SS, Bora RS, Al-Garni SM. Antimicrobial

activity of chitosan nanoparticles. Biotechnology & Biotechnological Equipment. 2021;**35**(1):1874-1880. DOI: 10.1080/13102818.2022.2027816

[104] Poznanski P, Hameed A, Orczyk W. Chitosan and chitosan nanoparticles: Parameters enhancing antifungal activity. Molecules. 2023;**28**(7):2996. DOI: 10.3390/molecules28072996

[105] El-Mohamedya RS, Abd El-Aziz ME, Kamel S. Antifungal activity of chitosan nanoparticles against some plant pathogenic fungi in vitro. Agricultural Engineering International: CIGR Journal. 2019;**21**:201-209

[106] Ing LY, Zin NM, Sarwar A, Katas H. Antifungal activity of chitosan nanoparticles and correlation with their physical properties. International Journal of Biomaterials. 2012;**2012**(1):632698. DOI: 10.1155/2012/632698

[107] Shih PY, Liao YT, Tseng YK, Deng FS, Lin CH. A potential antifungal effect of chitosan against *Candida albicans* is mediated via the inhibition of SAGA complex component expression and the subsequent alteration of cell surface integrity. Frontiers in Microbiology. 2019;**10**:602. DOI: 10.3389/fmicb.2019.00602

[108] Hassan EO, Shoala T, Attia AM, Badr OA, Mahmoud SY, Farrag ES, et al. Chitosan and nano-chitosan for management of *Harpophora maydis*: Approaches for investigating antifungal activity, pathogenicity, maize-resistant lines, and molecular diagnosis of plant infection. Journal of Fungi. 2022;**8**(5):509. DOI: 10.3390/jof8050509

[109] Xiaowei S, Chi Y, Liu T, Xun H, Yushou W, Liu X, et al. Black phosphorus Nanosheets-based hydrogel for efficient bacterial inhibition and accelerating wound healing. ACS Applied Materials & Interfaces. 2024;**16**(33):43892-43906. DOI: 10.1021/acsami.4c06075

[110] Buyuk NI, Arayici PP, Derman S, Mustafaeva Z, Yucel S. Synthesis of chitosan nanoparticles for controlled release of amiodarone. Indian Journal of Pharmaceutical Sciences. 2020;**82**(1):10-36468

[111] Jabbari N, Eftekhari Z, Roodbari NH, Parivar K. Evaluation of encapsulated eugenol by chitosan nanoparticles on the aggressive model of rheumatoid arthritis. International Immunopharmacology. 2020;**85**:106554. DOI: 10.1016/j.intimp.2020.106554

[112] Rawal T, Patel S, Butani S. Chitosan nanoparticles as a promising approach for pulmonary delivery of bedaquiline. European Journal of Pharmaceutical Sciences. 2018;**124**:273-287. DOI: 10.1016/j.ejps.2018.08.038

[113] Venkatesan KB, Alamelu S, Priya SR, Jayaseelan N, Kamaraj SK, Srinivasan MK, et al. Ameliorated antimicrobial, antioxidant, and anticancer properties by Plectranthus vettiveroides root extract-mediated green synthesis of chitosan nanoparticles. Green Processing and Synthesis. 2023;**12**(1):20230086

[114] Wong CY, Al-Salami H, Dass CR. Formulation and characterisation of insulin-loaded chitosan nanoparticles capable of inducing glucose uptake in skeletal muscle cells in vitro. Journal of Drug Delivery Science and Technology. 2020;**57**:101738. DOI: 10.1016/j.jddst.2020.101738

[115] Sharma M, Sharma R, Jain DK, Saraf A. Enhancement of oral bioavailability of poorly water soluble carvedilol by chitosan nanoparticles: Optimization and pharmacokinetic study. International Journal of Biological

Macromolecules. 2019;**135**:246-260. DOI: 10.1016/j.ijbiomac.2019.05.162

[116] Pauluk D, Padilha AK, Khalil NM, Mainardes RM. Chitosan-coated zein nanoparticles for oral delivery of resveratrol: Formation, characterization, stability, mucoadhesive properties and antioxidant activity. Food Hydrocolloids. 2019;**94**:411-417. DOI: 10.1016/j.foodhyd.2019.03.042

[117] Lin C, Kuo TC, Lin JC, Ho YC, Mi FL. Delivery of polysaccharides from *Ophiopogon japonicus* (OJPs) using OJPs/chitosan/whey protein co-assembled nanoparticles to treat defective intestinal epithelial tight junction barrier. International Journal of Biological Macromolecules. 2020;**160**:558-570. DOI: 10.1016/j.ijbiomac.2020.05.151

[118] Ling K, Wu H, Neish AS, Champion JA. Alginate/chitosan microparticles for gastric passage and intestinal release of therapeutic protein nanoparticles. Journal of Controlled Release. 2019;**295**:174-186. DOI: 10.1016/j.jconrel.2018.12.017

[119] He M, Zhong C, Hu H, Jin Y, Chen Y, Lou K, et al. Cyclodextrin/chitosan nanoparticles for oral ovalbumin delivery: Preparation, characterization and intestinal mucosal immunity in mice. Asian Journal of Pharmaceutical Sciences. 2019;**14**(2):193-203. DOI: 10.1016/j.ajps.2018.04.001

[120] Cole H, Bryan D, Lancaster L, Mawas F, Vllasaliu D. Chitosan nanoparticle antigen uptake in epithelial monolayers can predict mucosal but not systemic in vivo immune response by oral delivery. Carbohydrate polymers. 2018;**190**:248-254. DOI: 10.1016/j.carbpol.2018.02.084

[121] Saraf S, Jain S, Sahoo RN, Mallick S. Lipopolysaccharide derived alginate coated hepatitis B antigen loaded chitosan nanoparticles for oral mucosal immunization. International Journal of Biological Macromolecules. 2020;**154**:466-476. DOI: 10.1016/j.ijbiomac.2020.03.124

[122] Yu X, Wen T, Cao P, Shan L, Li L. Alginate-chitosan coated layered double hydroxide nanocomposites for enhanced oral vaccine delivery. Journal of Colloid and Interface Science. 2019;**556**:258-265. DOI: 10.1016/j.jcis.2019.08.027

[123] Dong W, Wang X, Liu C, Zhang X, Zhang X, Chen X, et al. Chitosan based polymer-lipid hybrid nanoparticles for oral delivery of enoxaparin. International Journal of Pharmaceutics. 2018;**547**(1-2):499-505. DOI: 10.1016/j.ijpharm.2018.05.076

[124] Hu Q, Hu S, Fleming E, Lee JY, Luo Y. Chitosan-caseinate-dextran ternary complex nanoparticles for potential oral delivery of astaxanthin with significantly improved bioactivity. International Journal of Biological Macromolecules. 2020;**151**:747-756. DOI: 10.1016/j.ijbiomac.2020.02.170

[125] Murthy A, Ravi PR, Kathuria H, Vats R. Self-assembled lecithin-chitosan nanoparticles improve the oral bioavailability and alter the pharmacokinetics of raloxifene. International Journal of Pharmaceutics. 2020;**588**:119731. DOI: 10.1016/j.ijpharm.2020.119731

[126] Yu A, Shi H, Liu H, Bao Z, Dai M, Lin D, et al. Mucoadhesive dexamethasone-glycol chitosan nanoparticles for ophthalmic drug delivery. International Journal of Pharmaceutics. 2020;**575**:118943. Available from: https://www.sciencedirect.com/science/article/pii/S0378517319309883

[127] Wang Y, Zhou L, Fang L, Cao F. Multifunctional carboxymethyl chitosan derivatives-layered double hydroxide hybrid nanocomposites for efficient drug delivery to the posterior segment of the eye. Acta Biomaterialia. 2020;**104**:104-114. DOI: 10.1016/j.actbio.2020.01.008

[128] Li J, Jin X, Yang Y, Zhang L, Liu R, Li Z. Trimethyl chitosan nanoparticles for ocular baicalein delivery: Preparation, optimization, in vitro evaluation, in vivo pharmacokinetic study and molecular dynamics simulation. International Journal of Biological Macromolecules. 2020;**156**:749-761. DOI: 10.1016/j.ijbiomac.2020.04.115

[129] Savin CL, Popa M, Delaite C, Costuleanu M, Costin D, Peptu CA. Chitosan grafted-poly (ethylene glycol) methacrylate nanoparticles as carrier for controlled release of bevacizumab. Materials Science and Engineering: C. 2019;**98**:843-860. DOI: 10.1016/j.msec.2019.01.036

[130] Shahab MS, Rizwanullah M, Alshehri S, Imam SS. Optimization to development of chitosan decorated polycaprolactone nanoparticles for improved ocular delivery of dorzolamide: In vitro, ex vivo and toxicity assessments. International Journal of Biological Macromolecules. 2020;**163**:2392-2404. DOI: 10.1016/j.ijbiomac.2020.09.185

[131] Wang H, Holgate J, Bartlett S, Islam N. Assessment of nicotine release from nicotine-loaded chitosan nanoparticles dry powder inhaler formulations via locomotor activity of C57BL/6 mice. European Journal of Pharmaceutics and Biopharmaceutics. 2020;**154**:175-185. DOI: 10.1016/j.ejpb.2020.07.011

[132] Ahmad N, Ahmad R, Alrasheed RA, Almatar HM, Al-Ramadan AS, Buheazah TM, et al. A chitosan-PLGA based catechin hydrate nanoparticles used in targeting of lungs and cancer treatment. Saudi Journal of Biological Sciences. 2020;**27**(9):2344-2357. DOI: 10.1016/j.sjbs.2020.05.023

[133] Vieira AC, Chaves LL, Pinheiro M, Lima SC, Neto PJ, Ferreira D, et al. Lipid nanoparticles coated with chitosan using a one-step association method to target rifampicin to alveolar macrophages. Carbohydrate Polymers. 2021;**252**:116978. DOI: 10.1016/j.carbpol.2020.116978

[134] Pardeshi CV, Agnihotri VV, Patil KY, Pardeshi SR, Surana SJ. Mannose-anchored N, N, N-trimethyl chitosan nanoparticles for pulmonary `administration of etofylline. International Journal of Biological Macromolecules. 2020;**165**:445-459. DOI: 10.1016/j.ijbiomac.2020.09.163

[135] Dhayanandamoorthy Y, Antoniraj MG, Kandregula CA, Kandasamy R. Aerosolized hyaluronic acid decorated, ferulic acid loaded chitosan nanoparticle: A promising asthma control strategy. International Journal of Pharmaceutics. 2020;**591**:119958. DOI: 10.1016/j.ijpharm.2020.119958

[136] Amiri N, Ajami S, Shahroodi A, Jannatabadi N, Darban SA, Bazzaz BS, et al. Teicoplanin-loaded chitosan-PEO nanofibers for local antibiotic delivery and wound healing. International Journal of Biological Macromolecules. 2020;**162**:645-656. DOI: 10.1016/j.ijbiomac.2020.06.195

[137] Fahimirad S, Abtahi H, Satei P, Ghaznavi-Rad E, Moslehi M, Ganji A. Wound healing performance of PCL/chitosan based electrospun nanofiber electrosprayed with curcumin loaded chitosan nanoparticles. Carbohydrate polymers. 2021;**259**:117640. DOI: 10.1016/j.carbpol.2021.117640

[138] Shafique M, Sohail M, Minhas MU, Khaliq T, Kousar M, Khan S, et al. Bio-functional hydrogel membranes loaded with chitosan nanoparticles for accelerated wound healing. International Journal of Biological Macromolecules. 2021;**170**:207-221. DOI: 10.1016/j.ijbiomac.2020.12.157

[139] Manne AA, Arigela B, Giduturi AK, Komaravolu RK, Mangamuri U, Poda S. Pterocarpus marsupium Roxburgh heartwood extract/chitosan nanoparticles loaded hydrogel as an innovative wound healing agent in the diabetic rat model. Materials Today Communications. 2021;**26**:101916. DOI: 10.1016/j.mtcomm.2020.101916

[140] AbdelAllah NH, Gaber Y, Rashed ME, Azmy AF, Abou-Taleb HA, AbdelGhani S. Alginate-coated chitosan nanoparticles act as effective adjuvant for hepatitis a vaccine in mice. International Journal of Biological Macromolecules. 2020;**152**:904-912. DOI: 10.1016/j.ijbiomac.2020.02.287

[141] Rebbouh F, Martin-Eauclaire MF, Laraba-Djebari F. Chitosan nanoparticles as a delivery platform for neurotoxin II from Androctonus australis hector scorpion venom: Assessment of toxicity and immunogenicity. Acta Tropica. 2020;**205**:105353. DOI: 10.1016/j.actatropica.2020.105353

[142] Abdelhamid HN, Dowaidar M, Langel Ü. Carbonized chitosan encapsulated hierarchical porous zeolitic imidazolate frameworks nanoparticles for gene delivery. Microporous and Mesoporous Materials. 2020;**302**:110200. DOI: 10.1016/j.micromeso.2020.110200

[143] Babii O, Wang Z, Liu G, Martinez EC, Chen L. Low molecular weight chitosan nanoparticles for CpG oligodeoxynucleotides delivery: Impact of molecular weight, degree of deacetylation, and mannosylation on intracellular uptake and cytokine induction. International Journal of Biological Macromolecules. 2020;**159**:46-56. DOI: 10.1016/j.ijbiomac.2020.05.048

[144] Rahmani S, Hakimi S, Esmaeily A, Samadi FY, Mortazavian E, Nazari M, et al. Novel chitosan based nanoparticles as gene delivery systems to cancerous and noncancerous cells. International Journal of Pharmaceutics. 2019;**560**:306-314. DOI: 10.1016/j.ijpharm.2019.02.016

[145] Mobarakeh VI, Modarressi MH, Rahimi P, Bolhassani A, Arefian E, Atyabi F, et al. Optimization of chitosan nanoparticles as an anti-HIV siRNA delivery vehicle. International Journal of Biological Macromolecules. 2019;**129**:305-315. DOI: 10.1016/j.ijbiomac.2019.02.036

[146] Baghaei M, Tekie FS, Khoshayand MR, Varshochian R, Hajiramezanali M, Kachousangi MJ, et al. Optimization of chitosan-based polyelectrolyte nanoparticles for gene delivery, using design of experiment: in vitro and in vivo study. Materials Science and Engineering: C. 2021;**118**:111036. DOI: 10.1016/j.msec.2020.111036

[147] Yan L, Gao S, Shui S, Liu S, Qu H, Liu C, et al. Small interfering RNA-loaded chitosan hydrochloride/carboxymethyl chitosan nanoparticles for ultrasound-triggered release to hamper colorectal cancer growth in vitro. International Journal of Biological Macromolecules. 2020;**162**:1303-1310. DOI: 10.1016/j.ijbiomac.2020.06.246

[148] Tang Y, Liu Y, Xie Y, Chen J, Dou Y. Apoptosis of A549 cells by small interfering RNA targeting survivin delivery using poly-β-amino ester/guanidinylated O-carboxymethyl chitosan nanoparticles. Asian Journal of Pharmaceutical Sciences.

2020;**15**(1):121-128. DOI: 10.1016/j.ajps.2018.09.009

[149] Nikkhoo A, Rostami N, Farhadi S, Esmaily M, Ardebili SM, Atyabi F, et al. Codelivery of STAT3 siRNA and BV6 by carboxymethyl dextran trimethyl chitosan nanoparticles suppresses cancer cell progression. International Journal of Pharmaceutics. 2020;**581**:119236. DOI: 10.1016/j.ijpharm.2020.119236

[150] Masjedi A, Ahmadi A, Atyabi F, Farhadi S, Irandoust M, Khazaei-Poul Y, et al. Silencing of IL-6 and STAT3 by siRNA loaded hyaluronate-N, N, N-trimethyl chitosan nanoparticles potently reduces cancer cell progression. International Journal of Biological Macromolecules. 2020;**149**:487-500. DOI: 10.1016/j.ijbiomac.2020.01.273

[151] Marques C, Som C, Schmutz M, Borges O, Borchard G. How the lack of chitosan characterization precludes implementation of the safe-by-design concept. Frontiers in Bioengineering and Biotechnology. 2020;**8**:165. DOI: 10.3389/fbioe.2020.00165

[152] El-Kamary SS, Pasetti MF, Mendelman PM, Frey SE, Bernstein DI, Treanor JJ, et al. Adjuvanted intranasal Norwalk virus-like particle vaccine elicits antibodies and antibody-secreting cells that express homing receptors for mucosal and peripheral lymphoid tissues. The Journal of Infectious Diseases. 2010;**202**(11):1649-1658. DOI: 10.1086/657087

[153] Mann AJ, Noulin N, Catchpole A, Stittelaar KJ, De Waal L, Veldhuis Kroeze EJ, et al. Intranasal H5N1 vaccines, adjuvanted with chitosan derivatives, protect ferrets against highly pathogenic influenza intranasal and intratracheal challenge. PLoS One. 2014;**9**(5):e93761. DOI: 10.1371/journal.pone.0093761

Chapter 2

Chitosan-Based Materials for Effective Wound Healing

Isha Murtaza, Muhammad Yar, Zulfiqar Ali Khan and Sohail Anjum Shahzad

Abstract

Wound healing is a complex biological process calling for materials that improve tissue regeneration, control infections, and sustain a favorable environment for healing. Biopolymers are proven to be more effective than synthetic ones because of their exceptional biodegradability, biocompatibility, and antimicrobial properties. Chitosan, a naturally derived polymer, due to its antioxidant, regenerative, and antimicrobial properties, has emerged as a versatile candidate for wound healing applications. This chapter explores the versatile uses of chitosan-based materials in wound management, involving hydrogels, nanocomposites, films, sponges, and membranes. It emphasizes the synthesis, biological evaluation, and formulation methods of these materials and their role in promoting different phases of wound healing. By incorporating bioactive agents, nanoparticles, and synergistic polymers into chitosan matrices to improve the antimicrobial activity, mechanical strength, and therapeutic efficacy. This chapter highlights the potential of chitosan-based materials for effective wound healing and addresses both the opportunities and challenges in their clinical translation.

Keywords: chitosan-based materials, wound dressing, hydrogels, membranes, nanocomposites, films, sponges

1. Introduction

The skin, the largest exposed organ of the body, encompasses around 8% of total body mass and prevents the body from physical harm and microbial intrusion [1]. The skin has a multilayered structure, and each layer serves a distinct function. The outer epidermis, mainly composed of keratinocytes, acts as a protective barrier against pathogens and UV radiation. Below it is the dermis, a dense connective tissue layer rich in collagen that provides elasticity, mechanical strength, and nutrition to the epidermis; also, fibroblasts produce elastin and collagen. The dermis and epidermis coordinate to maintain hemostasis. The hypodermis is the innermost layer, consisting of adipose tissue, offering mechanical protection and thermal insulation [2].

Wounds are any damage or disorder in the skin's healthy structure and function, owing to trauma, genetic disorder, surgical intervention, and injury, as well as scalding, surgery, and chronic ulcers [1, 3]. Wounds are categorized as acute or chronic based on their healing time. Acute wounds that result from sudden injuries normally heal within 2–3 months, depending on their severity. Chronic wounds, such as ulcers, burns, and infections that are more severe and do not heal within the expected timeframe, could lead to serious health risks [2, 4, 5]. Wounds are classified as superficial, partial-thickness, or full-thickness on the basis of injury depth [1].

When the skin barrier is compromised, it provides an ideal environment for both aerobic and anaerobic bacterial infections [6]. Wound healing is a complex, coordinated process that occurs in sequential phases: hemostasis, inflammation, proliferation, and remodeling. These phases are firmly regulated in both time and space to ensure effective healing and tissue repair [2, 7].

Hemostasis initiates immediately after skin injury [8]. During this phase, blood vessels constrict, and platelets are activated upon contact with exposed collagen, releasing granules that promote further platelet activation and aggregation. This, when combined with the coagulation cascade, leads to the formation of a provisional fibrin matrix within the wound [9]. The platelet activation releases cytokines like TGF-β and platelet-derived growth factor, triggering neutrophil and macrophage recruitment to initiate the inflammatory phase. The neutrophils are dispensable, but macrophages are crucial for phagocytosis and secrete factors that promote fibroblast proliferation, angiogenesis, and keratinocyte migration. If the macrophage function is deregulated, it impairs wound healing, especially in diabetic wounds [2].

The proliferative phase activates after 2–3 days of injury and continues up to 3 weeks in cutaneous wound healing. In this phase, fibroblasts migrate to the wound, producing disorganized collagen, mainly immature type III collagen, within the provisional matrix. Then, some fibroblasts transform into myofibroblasts under cytokine influence, enhancing collagen production and facilitating wound contraction. Then the remodeling phase begins and replaces granulation tissue with a permanent scar. Active collagen production continues for 4–5 weeks, followed by the replacement of type III collagen with type I fibrillar collagen over a year. The matrix metalloproteinases and zinc-dependent enzymes are secreted by epidermal cells that are crucial for tissue remodeling. Collagen production improves a wound's tensile strength, increasing from 3% at week 1 to 20% by week 3. After 3 months of injury, the tensile strength peaks at 80% of undamaged skin but never fully restores to 100%. Each phase of wound healing is crucial for the successful restoration of tissue function and integrity [9].

Wound management is a major global challenge that costs billions annually and drives focus on the development of innovative wound dressings [10]. Wound dressings are the materials designed to cover wounds, provide protection, and facilitate the healing process [11]. Wound dressings play an important role in supporting the healing process by covering the wound and maintaining an optimal environment. They have properties like fluid absorption, elasticity, moisture retention, biocompatibility, stimulation of growth factors, and protection against secondary contamination. Moreover, dressings must have oxygen permeability, nontoxicity, a porous structure, barrier effectiveness, and antibacterial properties [12, 13].

Dressings are classified based on their function (debridement, antibacterial, occlusive, absorbent, adherence), material (hydrocolloid, alginate, collagen), and physical form (ointment, film, foam, gel). They are further characterized into primary dressings that directly contact the wound and secondary dressings that cover the primary dressing and island dressings, which have an absorbent core surrounded by adhesive. Further classifications involve traditional, modern, and advanced dressings, skin replacement products, and wound healing procedures [14]. An ideal wound dressing should exhibit the following attributes: (1) nontoxicity and non-irritation,

(2) biodegradability in vivo, (3) strong antibacterial properties to inhibit infections, (4) optimal moisture and air permeability to minimize dressing changes, and (5) adequate mechanical strength to resist wrinkling [15].

Conventional wound dressings, even though cost-effective with good exudate absorption, have limited efficacy in wound healing. To address this, natural polymers like alginate, hyaluronic acid, cellulose, collagen, and chitosan, alongside synthetic polymers like poly(vinyl alcohol) and polyurethane, are used in advanced dressings. Natural polymers excel in biochemical properties, while synthetic ones offer superior mechanical performance [2].

From all the biopolymers, chitosan is especially notable for its biodegradability, biocompatibility, antibacterial, and antioxidant properties [2]. Chitosan and its derivatives contribute significantly to the first three phases of wound healing. In the hemostasis phase, they promote platelet aggregation and hinder fibrin dissolution. Throughout the inflammation phase, they help remove bacteria from the wound. In the proliferation phase, chitosan accelerates granulation tissue growth and facilitates skin regeneration. These actions support the overall healing process, culminating in tissue remodeling [3].

2. Chitosan-based biomaterial for wound healing and tissue engineering

Tissue engineering is an advanced field of medicine that combines cells, biologically active molecules, and scaffolds to restore, improve, and conserve the function of injured tissue or organs [16]. Various materials have been developed to suit specific wound types and healing mechanisms. These include synthetic polymers like silicone and polyurethane rubber, as well as biopolymers such as collagen, chitosan, gelatin, and alginate [17].

Recently, biologically active natural materials have gained significant attention for tissue engineering due to their unique ability to mimic human tissue structure. Their physical and chemical resemblance to natural tissues, combined with lower toxicity and superior biocompatibility compared to synthetic polymers, has driven their demand over the past two decades. Despite advancements, repairing and regenerating injured tissue remains a clinical challenge, necessitating the development of innovative and robust biomaterials. Both natural and synthetic materials, including chitosan, collagen, gelatin, alginate, silk fibroin, hydroxyapatite, hyaluronic acid, polyethylene glycol, polylactic acid, poly(lactic-co-glycolic) acid, and polycaprolactone, have been explored for tissue engineering applications [18].

Among biopolymers, chitosan has garnered significant attention as a promising bioactive material due to its unique properties and abundance. In recent years, it has demonstrated exceptional potential in tissue engineering applications, including bone, skin, cartilage, intervertebral discs, and blood vessels, making it highly effective for repairing diseased and damaged tissues [19]. Chitosan, a polysaccharide, is extensively applied in biomedical applications. It is derived from the partial deacetylation of chitin (**Figure 1**), the second most abundant polysaccharide, commonly found in the exoskeletons of crustaceans and insects, as well as in fungal cell walls [20].

Chitosan holds a distinct position among biomaterials due to its abundance, versatility, and remarkable properties, such as biodegradability, biocompatibility, nontoxicity, hydrophilicity, antibacterial and antifungal activities, and wound-healing efficacy [21]. Additionally, the presence of β-(1,4) glycosidic bonds between D-glucosamine and N-acetyl-D-glucosamine allows for easy chemical modifications

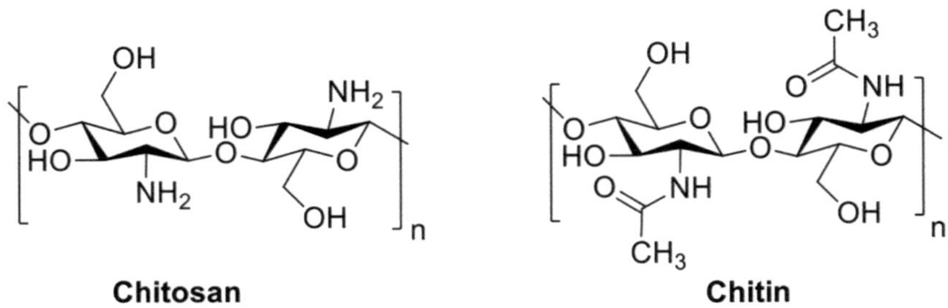

Figure 1.
Structure of chitosan and chitin [3].

of chitosan, enhancing its elasticity and flexibility and reducing inflammatory responses [14].

Furthermore, the degradation of chitosan yields harmless amino sugars that are fully absorbable by the human body. These attributes establish chitosan as a versatile candidate for a wide range of biomedical applications, including drug delivery systems, surgical sutures, wound healing, and tissue engineering [22, 23]. Due to its brittleness, chitosan films exhibit weak elasticity and tensile strength. To enhance these properties, chitosan is often cross-linked with other polymers, improving its elasticity and tensile strength for better performance in biomedical applications [24].

Chitosan-based materials have been widely utilized in various biomedical applications (**Table 1**), including wound healing [20]. These materials can be processed into different forms such as sponges, hydrogels, nanocomposites, membranes, beads, and scaffolds, each offering unique advantages for promoting tissue repair and regeneration. Chitosan and its derivatives are being explored for biomedical and pharmaceutical applications in drug delivery, tissue engineering, and wound healing due to their bacteriostatic effects, rejuvenative ability, and biodegradability. They improve mineralization, collagen production, tissue regeneration, osteoblast activity, and hemostatic action [25].

Biopolymer wound dressing type	Specification	Results
Hydrogels	Physically associated (reversible) Chemically cross-linked (irreversible)	Enhanced wound healing due to rat skin fibroblast growth (in vitro research).
Sponges	Prevent infection	A sponge loaded with antibiotics can be used as a medicine carrier to treat wounds and prevent infections.
Membranes	High concentration Increased stability Excellent oxygen permeability	Improves tissue regeneration and decreases re-epithelialization in human subjects (clinical research).
Films	Good adhesion Free standing Improved mechanical and swelling properties	Three different types of cytokines are secreted, adhesion is enhanced, and proliferation is promoted by chitosan types with a low degree of deacetylation (in vivo research).

Table 1.
Selected preclinical and clinical research on chitosan as a biomaterial for wound dressings [14].

Chitosan-based materials show significant potential in wound healing due to biodegradability, biocompatibility, and intrinsic antimicrobial properties. However, there are still several obstacles to overcome in their clinical translation from laboratory products, including regulatory approval, expensive clinical studies, industrial scalability, wound kinds, and unique patient needs. The mechanical properties of chitosan, like tensile strength and elasticity, remain suboptimal and need improvement by cross-linking or composite formulations. Furthermore, coupling with smart biomaterials and nanotechnology can help with better controlled drug delivery and bioactivity. Material properties and therapeutic effects are highly diverse due to variations in chitosan's molecular weight, origin, processing techniques, and degree of deacetylation. Active natural metabolites incorporated into chitosan formulations have great wound healing effects, but their use in the clinics is hampered by toxicity and insufficient in vivo studies. Future research must address these issues by modifying chitosan's characteristics for precise tissue requirements or wound types to improve their efficacy and safety [14]. There is a need for a multidisciplinary approach integrating researchers, clinicians, and industrial partners to prepare chitosan-based scaffolds that mimic human tissues for effective wound healing. To ensure the safe and effective use of these biomaterials in clinical practice, there is a dire need to standardize manufacturing and comprehensive clinical validation. The full therapeutic potential of chitosan could be unlocked by coping with these challenges, establishing it as a foundation in advanced wound management.

2.1 Chitosan-based hydrogels

Hydrogels are composed of a solid phase and a liquid phase, predominantly water. Because the solid phase gives the structure to gel, it can absorb great volumes of water without becoming soluble. Due to their high water content, softness, flexibility, and biocompatibility, hydrogels are distinguished biomaterials that reduce tissue damage throughout implantation. Hydrogels are frequently employed as biomedical scaffolds for tissue replacement and in applications such as the transport of growth factors and drugs. They could assist in tissue repair both morphologically and functionally because of their mechanical characteristics, which are very similar to those of soft body tissues [26].

Because of their distinct physiochemical characteristics, hydrogels that are made up of networks of hydrophilic polymers created by chemical or physical interactions are being investigated extensively in biomedicine. They are favorable wound dressings because of their high water absorption capacity that permits them to absorb tissue fluids from wounds. Moreover, hydrogels' porous nature assists the stabilization and dispersion of nanoparticles, making them proficient carriers for these substances. Chitosan-based hydrogels are unique due to their biodegradability, cost-effectiveness, nontoxicity, biocompatibility, hemostatic qualities, and intrinsic antibacterial activity. Through Van der Waals forces, electrostatic forces, and hydrophobic interactions, the amino groups in chitosan involve nonselectively with both Gram-positive and Gram-negative bacteria, facilitating bacterial aggregation and isolation [27].

Chopra et al. [4] developed chitosan/polyvinyl alcohol (PVA)-based hydrogel films loaded with honey using a solvent-casting method for potential wound healing applications. To increase their biocompatibility, physically cross-linked hydrogels were created without using harmful organic solvents. These films were prepared by combining chitosan and polyvinyl alcohol (PVA) solutions with honey in varying proportions, followed by air drying. These hydrogels showed notable swelling, moisture uptake, and better mechanical strength, all of which are necessary characteristics for wound dressings. The surface morphology and molecular interactions of these hydrogels were studied by advanced techniques such as scanning electron microscopy (SEM), FT-IR, and differential scanning calorimetry (DSC). The antibacterial activity of these hydrogel films was studied against *Staphylococcus aureus*. They showed a great antibacterial effect, being bacteriostatic. The prepared hydrogel films exhibit regulated in vitro honey release, interaction between honey's components and wound healing proteins, and physical stability, considering them as efficient and biocompatible wound dressings [4].

Biopolymer hydrogels containing cellulose and chitosan showed great properties: high porosity, antibacterial efficiency, and controlled drug release crucial for wound management. When cross-linked with synthetic polymers, these hydrogels demonstrated superior mechanical properties, like elasticity, flexibility, and tensile strength, which are essential characteristics of effective wound dressings. Wound healing was enhanced through the incorporation of antimicrobial compounds and metallic nanoparticles in hydrogels by re-epithelialization, granulation tissue growth, and inhibition of pathogens. Persistent antimicrobial activity was obtained by incorporating nanotechnology while normal cell viability was not compromised. Though the in vivo and in vitro results of hydrogels were satisfactory, more innovations and preclinical testing of chitosan and cellulose-based hydrogels are needed to validate their wound healing applications, efficacy, and safety for clinical applications [5].

Zhang et al. [28] used aqueous emulsion copolymerization to synthesize a novel hydrogel film developed from carboxymethyl chitosan (CMCS) loaded with waterborne polyurethane-gelatin hydrolysate (WPU-g-GH). Superior dry/wet mechanical strength, biodegradability, and water absorption are shown by these hydrogel films, particularly those with 6% CMCS. Strong hydrogen bonding between CMCS and WPU-g-GH molecules and a uniform, smooth film surface was characterized by X-ray photoelectron spectroscopy (XPS), FT-IR, and SEM. They showed significant antibacterial activity against *Escherichia coli* and *Staphylococcus aureus*, making them ideal for wound healing applications [28].

Jung et al. [29] prepared chitosan/polyethylene glycol (CH/PT) hybrid hydrogels by mixing the polymer solutions in situ, then cross-linked them by hydrogen peroxide (H_2O_2) and horseradish peroxidase (HRP). By varying the molecular weight, content, and configuration of PEG, these hydrogels showed adjustable properties. The prepared hydrogels exhibited great mechanical strength, involving toughness (45 J/m^3), compressive strength (4.9 MPa), and adhesive strength (192 kPa), considering them desirable wound dressings. They have proved to be hemostatic agents when applied on defected liver models of rats and mice. These hydrogels also demonstrated enhanced biodegradability, quick gelation, and the capability to regulate degradation rates. These CH/PT hydrogels show great potential for use in wound healing, tissue engineering, and other applications needing strong tissue adhesiveness and precise mechanical properties [29].

Xia et al. [30] developed carboxymethyl chitosan-oxidized hyaluronic acid (CMCS-OHA), a biodegradable chitosan-based hemostatic hydrogel, and evaluated its hemostatic activity in a rat liver injury model. The hemostatic efficacy of prepared hydrogel was found comparable to that of fibrin sealant (FS), with an analogous hemostatic time and blood loss. The CMCS-OHA improved post-surgery wound healing and showed negligible inflammatory response. These CMCS-OHA hydrogels were considered biocompatible as they were absorbed and excreted quickly through urine without accumulating in the tissues, as revealed by biodegradation studies. These hydrogels could be clinically used as a hemostatic sealant due to their biosafety, biodegradability, biodistribution, urinary excretion, and ability to control bleeding [30].

Chitosan-based hydrogels prepared with aloe vera juice were synthesized by Drabczyk et al. [31] through UV irradiation, with the purpose of increasing the therapeutic efficacy for wound healing purposes. The hydrogels exhibited high biocompatibility, as evidenced by more than 95% viability of murine fibroblasts, as well as good swelling capacity, which are crucial for absorbing wound exudates and providing a humid healing environment. The addition of aloe vera juice enhanced the sorption behavior of the hydrogels by virtue of its hydrophilicity and also led to a more homogeneous surface morphology and greater mechanical ductility. These hydrogels have tunable physicochemical properties in the range of amounts and distribution of the crosslinker used and contain aloe vera, so they are desirable for biomedical applications such as wound dressing [31].

Shamloo et al. [32] made chitosan hydrogels with honey, gelatin, and polyvinyl alcohol (PVA) using the freeze-thawing method for wound healing. The hydrogels were biocompatible, antibacterial, and had tunable mechanical properties. Honey added antibacterial activity and cell proliferation. In vivo studies on rat models showed that honey-containing hydrogels (10% and 20%) accelerated wound healing, collagen deposition, fibroblast proliferation, and angiogenesis. The hydrogels created a moist wound environment, reduced inflammation, and transitioned the wound through the healing phases. Chitosan hydrogels are promising for advanced wound care [32].

2.2 Chitosan-based nanocomposite matrix

Chitosan, a biopolymer, has been an attractive nanocomposite matrix due to its biocompatibility, biodegradability, and versatile chemical properties. The incorporation of various types of nanoparticles, such as inorganic, organic, and metallic structures, greatly enhances the mechanical, thermal, and chemical robustness of chitosan. These nanocomposites are widely utilized in applications including wastewater treatment, where chitosan adsorbs pollutants efficiently, such as heavy metals and dyes; biomedical applications for drug delivery systems and tissue engineering as well as corrosion inhibition applications. The fabrication technology improvement by using electrospinning or 3D printing yields a more precise structural control on the size for niche applications. However, further improvements have to be made in order to cater to large-scale production for industrial usage; meanwhile, expansion toward other sectors, including energy storage and food packaging, still has much space for exploration [33].

Sakthiguru et al. [34] prepared zinc oxide (ZnO) nanoparticles, O-carboxymethyl chitosan (O-CMC), and Lawson (Ln) nanocomposites by ex-situ grafting to improve wound healing. Advanced techniques such as SEM, TGA, FT-IR, TEM, and X-ray diffraction (XRD) were used for the characterization of developed nanocomposites. They revealed improved thermal stability, structural integrity, and functional group confirmation. The MTT assay was used to evaluate biocompatibility, and L929 fibroblast cells were used to examine wound scratch assays, demonstrating great anti-inflammatory, antioxidant, and antibacterial properties. The prepared nanocomposites incorporate the antimicrobial effect of ZnO, bioactivity and solubility of O-CMC, and therapeutic power of Ln, with sustained release in Dulbecco's minimal essential medium (DMEM) and phosphate buffer saline (PBS), superior drug encapsulation, and effective cell migration capabilities. So, nanocomposites based on ZnO, O-CMC, and Ln have the power to be employed as essential materials for tissue engineering and wound healing [34].

Shah et al. [35] developed chitosan-sericin-silver nanocomposite (CSSN) films both with and without moxifloxacin (Mox) by using in situ reduction of silver nanoparticles in a chitosan-sericin colloidal matrix, and then solvent casting was applied. Nanoparticle analysis was done by zeta size and UV-visible spectroscopy, film properties by energy-dispersive X-ray spectroscopy (EDX), scanning electron microscopy (SEM), X-ray diffraction (XRD), fourier transform infrared spectroscopy (FT-IR), differential scanning calorimetry (DSC), and thermogravimetric analysis (TGA) techniques, examined for content uniformity, swelling behavior, permeation, and Mox release. The films exhibited sustained drug release (up to 78.57% in 24 hours) and high antibacterial activity against multiple pathogens. Enhanced wound healing with better collagen reorganization, fibrosis, epidermal regeneration, and neovascularization, without silver ion toxicity, was demonstrated by in vivo testing on burn-injured rats. These qualities, combined with nanocomposites' mechanical stability, hydrophilicity, and biocompatibility, give them promising properties for good wound dressing [35].

Nowroozi et al. [36] investigated the development of chitosan (CS), curcumin (Cur), and graphene oxide (GO) 3D scaffolds for wound healing applications using the freeze-drying method. Hammer's method was used to synthesize graphene oxide, and its antibacterial properties were improved by loading it with curcumin. The prepared scaffolds were examined for antibacterial effects, mechanical strength, hydrophilicity, and cell viability. It showed that the CS matrix incorporated with GO/Cur reduced pore size, improved mechanical properties, and enhanced phosphate-buffered saline (PBS) absorption, with the 1% GO/Cur scaffold showing optimal performance. A notable reduction in bacterial growth was validated by antibacterial tests, and almost 95% viability was indicated by cell viability assays. The antibacterial, biological, and structural characteristics collectively employ the CS/GO/Cur scaffolds as effective wound dressing material [36].

Azadmanesh et al. [37] fabricated novel 3D-printed wound healing scaffolds using hyaluronic acid (HA), poly-lactic acid (PLA), rosmarinic acid, chitosan, and copper carbon dots (Cu-CDs). Hydrothermal methods used to synthesize Cu-CDs showed great antibacterial activity against Gram-positive bacteria, especially at the concentration of 1 mg/ml. Good biocompatibility was shown by the scaffold, and desirable results were obtained for cytotoxicity tests on L929 fibroblast stem cells. The PLA/HA/Cu-CDs/rosmarinic acid/chitosan composite enhanced wound healing as compared to the only PLA scaffold, demonstrated by in vivo experiments on rats. Improved tissue regeneration was confirmed by gene expression data (PDGF, TGF-β, and MMP-1) and histological analysis. By combining biocompatibility, tissue-regenerative capabilities, and antibacterial effects, these scaffolds can be a promising approach to treat chronic wounds [37].

Kędzierska et al. [38] examined the wound healing applications of cellulose-filled chitosan nanocomposite films, emphasizing their biological, hemostatic, and mechanical properties. Nanosized cellulose (CNC) and microcrystalline cellulose (MCC) were used to secure these films and then reformed with phosphorylated derivatives (P-CNC and P-MCC). Improved hemostatic activity is shown by these films with a variation of both coagulation pathways to drop clotting times due to the addition of cellulose fillers. Furthermore, these films exhibited enhanced skin cell migration necessary for wound healing with negligible cytotoxicity, no genotoxicity, and antioxidant qualities. All these properties proposed the use of these biocompatible, biodegradable, and transparent nanocomposite films as innovative wound dressings, especially for chronic wounds [38].

Biomedical uses such as drug delivery, tissue engineering, and wound healing are promising for chitosan-based nanocomposite scaffolds. Mohandas et al. [39] found that chitosan matrices enhanced antimicrobial properties and were cytocompatible with metallic nanoparticles, including silver, gold, copper, zinc oxide, and titanium dioxide. The problems of wound management, like infection, exudate management, and inflammation, were targeted in this way. The metallic ions helped in wound healing since they aided in the processes of epithelial cover and affected inflammation. Further investigation was necessary due to their cytotoxicity and environmental consequences. However, striking preclinical and clinical trials are essential for translating these advancements into clinical practice, which will inevitably rely on the development of optimized scaffold designs to suit particular wound conditions [39].

The novel combinations of chlorhexidine, chitosan glycolate, and chitin nanofibrils were prepared by Muzzarelli et al. [40] in three forms: spray, gauze, and gel. Gauze was assimilated with biocompatible reinforcement of non-woven dibutyryl chitin. They used phytostimuline-medicated wounds as control and carried out two series of tests in a murine wound model to examine the prepared product. One test involved the treatment of concomitant laser that could initiate cells. The gauze continuously promoted epithelial differentiation, keratinization, and basal lamina reformation. Regardless of laser treatment, the gauze prompted an angiogenesis reaction that was used to investigate vascular trophism. Immunohistochemical observations showed the development of the actin cytoskeleton of epidermal and dermal cells. The gauze was used to treat the traumatic wounds of 75 hospitalized patients and showed satisfactory results. In conclusion, the gauze restored a scarless and physiological epidermis, so it was considered a better dressing. Meanwhile, the gel was recommended for thin layer areas as it improved physiological repair, and the spray can be used as a first aid treatment for bleeding abrasions. To support their commercial viability with respect to healing time and medical work rate, the biochemical significance of the chitin and chitosan materials was determined [40].

Quaternized chitosan nanocomposite films with silver nanoparticles intended for antimicrobial wound dressing application were prepared by Rahimi et al. [41]. The films were characterized using FT-IR, XRD, SEM, and EDX techniques. It was observed that the antimicrobial action of the films was considerably enhanced by incorporating silver nanoparticles. The results of swelling tests indicated that the swelling ratio of the nanocomposites was decreased when a higher content of silver nanoparticles was used, which was attributed to greater matrix density and less hydrophilic sites. Hemolysis and blood clotting studies showed that the films developed were substantially fewer hemolytic and possessed a better clotting ability even at low silver content, which indicated good blood compatibility. Antimicrobial assays proved that even pure quaternized chitosan films (silver-free) contain antimicrobial activity due to the presence of imidazolium functional groups and were able to suppress *Escherichia coli*, *Staphylococcus aureus*, *Pseudomonas aeruginosa*, and *Candida albicans* pathogens. The enhancement of antimicrobial activity against multidrug-resistant *P. aeruginosa* was achieved with the incorporation of silver nanoparticles. Cytotoxic and cell attachment tests revealed that the modified films containing a lower dosage of silver are highly biocompatible and can promote human fibroblast cell growth. Such results indicate that quaternized chitosan nanocomposites hold great promise for advanced antimicrobial wound dressings [41].

Dutta et al. [42] prepared and characterized chitosan-sepiolite (CS-SEP) nanocomposite films using the solution casting method for wound healing applications. Physicochemical and biological properties of these films were evaluated by FT-IR,

SEM, and XRD techniques. Results showed that the addition of sepiolite increased water absorption, water contact angle, folding endurance, porosity, antibacterial activity, and blood clotting ability of the films, making them more suitable for wound care. Optimized CS7SEP3 nanocomposite showed excellent antibacterial activity against *Bacillus subtilis* and *Escherichia coli*, non-hemolytic behavior, and high cell viability and proliferation in vitro. These results suggest that CS-SEP nanocomposite films are good materials for wound management with improved mechanical strength, moisture retention, and biological compatibility than pure chitosan films [42].

In order to treat chronic wounds in patients, Sandri et al. [43] combined chitosan oligosaccharide (COS) and halloysite nanotubes (HNTs) into a new type of nanocomposite. This nanocomposite was fabricated via ionic interactions by using HNTs as a carrier of COS. The hybrid material showed good biocompatibility, which allowed fibroblasts to proliferate and migrate in vitro. HNTs/COS nanocomposite significantly improved the rate of wound healing in the murine model by encouraging early epithelial formation, blood vessel formation, and granulation tissue development. Based on the apparent microscopic structure of the tissues, collagen had been fairly organized inside the wounds even after 18 days of full skin healing. The results indicate that HNTs/COS nanocomposites have great potential as an efficient and suitable material for wound healing and tissue repair enhancement [43].

Shanmugapriya et al. [44] established a multifunctional nanocomposite film linking bentonite (BN), polyvinylpyrrolidone (PVP), and chitosan (CS) for innovative wound care that possessed better physical properties, biocompatibility, and antibacterial activity. It was observed that the strong bonding or interactions between CS, PVP, and BN enhance the thermal stability, tensile strength, porosity, and water absorption capacity of the film, thus making it very conducive to a moist healing environment. In vitro cell culture cytotoxicity studies against L929 and NIH3T3 fibroblast cells showed no observable toxicity, while conversely, good in vivo results of 97% wound closure within 16 days, faster healing, and less scarring than controls were also reported. It was concluded that there were significant increases in collagen deposition, re-epithelialization, and fibroblast proliferation as evidenced in the histological analysis, thereby proving that the material has potential as a more advanced wound dressing. The ability of the CS-PVP-BN2 variant to exhibit greater antibacterial activity, mechanical strength, and controlled degradation made it a suitable biomaterial for wound care products [44].

Chitosan-fibrin nanocomposites (CFNs) were developed by Vedakumari et al. [45], adopting a new co-precipitation technique, and obtained nanoparticles with a 24–28 nm diameter and with a zeta potential of +16 mV. These composites displayed good biocompatibility, antibacterial activity against *Escherichia coli* and *Staphylococcus aureus*, as well as improved thermal stability due to chitosan and fibrin interaction. CFNs loaded with methotrexate (MTX) maintained their release for 96 hours and showed a cytotoxic effect on both HeLa and MCF-7 cancer cell lines in a dose-dependent manner. In vivo studies done using albino rats showed a speedy curative process as the wounds of the albino rats were healed completely within 14 days compared to the control, where it took 22 days. CFN treatment improved collagen production, migration of the fibroblast, and tensile strength from both biochemical and histological studies. Given the findings, CFNs appear to be a potential reach in active pharmaceutical ingredients (API) nanocomposite material useful in wound healing with inherent antibacterial activity and biocompatible properties [45].

Conventional wound bandages have certain drawbacks, including stiffness, poor mechanical properties, non-permeability, and adhesion to wounds. Also, most of them do not provide bactericidal protection, which is important for wound healing. To meet these challenges, Zhai et al. [46] found that the introduction of a nanocomposite, namely keratin-chitosan with nano-ZnO, has yielded better results in terms of swelling, bactericidal activity, and biocompatibility. The nanocomposite was also found to support better wound healing, as seen by increased collagen deposition and skin tissue formation in the rat models. Also, the use of zinc oxide nanoparticles provided antimicrobial properties, thus protecting the wounds from getting infected, which is very vital in wound management. These nanocomposites possessed high mechanical strength, biodegradability, and efficient antimicrobial activity; they have tremendous potential to be used as an innovative dressing material for burn wounds [46].

2.3 Chitosan-based membranes

Chitosan-based materials have many applications in biomedical processes, such as wound healing and drug delivery. Chitosan and its derivatives have been shown to inhibit biofilm formation and reduce bacterial virulence, thus making them potentially useful in medical treatments. But their antibacterial activity can be enhanced by combining them with other agents. Recently, chitosan-based composite membranes for wound healing have been developed, which have improved antibacterial properties, promote cell adhesion, and accelerate tissue repair. These composite materials often contain other biopolymers like curcumin, aloe vera, and quaternary ammonium compounds to enhance their healing and antimicrobial properties. Chitosan has also been used in drug delivery systems, such as nanoparticle-based formulations for targeted therapy and in gene delivery systems for cancer treatment. Moreover, chitosan membranes have been used in various catalytic processes, fuel cells, and environmental applications where they have shown promise in wastewater treatment, heavy metal adsorption, and energy generation. The ongoing research is focused on modifying the material properties by combining them with other compounds to improve stability, functionality, and applicability in different fields [47].

Chitosan-based electrospun membranes have been shown to be highly promising in wound healing applications because of their biodegradability, biocompatibility, antimicrobial activity, and ability to provide a moist healing microenvironment. These membranes that are commonly applied to burn and chronic wound dressings trapped wound exudate, accelerated tissue regrowth, and decreased rates of infection. Nevertheless, issues like batch-to-batch heterogeneities, mechanical weakness, and the need for enhanced cell infiltration and vascularization limited their clinical translation. Recent developments like the integration of growth factors, nanoparticles, and advanced electrospinning have the potential to be effective. Despite these advancements, additional studies are needed to further tailor the biodegradation, mechanical strength, and biological activity for wider clinical applications of chitosan membranes [48].

Fibroblast growth and microbial infection are two main elements that influence wound healing. The differentiation, proliferation, and survival of fibroblasts are sustained by the initiation of extracellular matrix by natural polymer dressing membranes. They developed the composite membranes of chitosan/titanium with TiO_2 in varying percentages. Mechanical properties, thermal behavior, ultrastructure, and morphology of membranes were characterized by tensile strength, SEM, XRD, and FT-IR techniques. Strong O–Ti–O bonding between chitosan and titanium oxide improved the flexibility, porosity, crystallinity, and mechanical strength of membranes. The membrane with 0.25% TiO_2 reduced apoptosis and oxidative stress and increased survival and proliferation in L929 cells. Moreover, the membrane with 0.25% showed good antibacterial activity against *Staphylococcus aureus*. All the results indicated that the membrane with 0.25% has good functional integrity, survival, and fibroblast growth and good antibacterial activity, so they can be used as wound dressing [49].

Meng et al. [50] prepared blended, drug-loaded chitosan membranes using a casting/solvent evaporation method. The membranes were made from different ratios of alginate and chitosan with silver sulfadiazine as a model drug. The polyblend composition of 1:1 showed the highest viscosity. The alginate/chitosan membranes had a water vapor transmission rate between 442 and 618 $g/m^2/day$; water uptake properties were affected by ionic strength and pH. The wet membrane's tensile elongation was 46.28%, and the dry membrane's tensile strength was 52.16 MPa. The controlled release study showed that when the alginate content was 50%, silver sulfadiazine was released faster. These membranes have good physical properties, pH-dependent water uptake, and controlled release behavior, and they are biodegradable and suitable for wound healing applications [50].

Lin et al. [51] developed zwitterionic chitosan soft membranes for diabetic wound healing by implanting a copolymer of glycidyl methacrylate (GMA) and sulfobetaine methacrylate (SBMA). FT-IR and XPS revealed that the copolymer was successfully grafted onto the chitosan matrix with quaternary ammonium and sulfonate groups. The SEM images showed a highly porous structure for exudate drainage and gas exchange. Hydration properties (water contact angle and swelling ratio) demonstrated that the grafted membranes were more hydrophilic. Biocompatibility tests showed reduced cell and bacteria adhesion; G20S80 was the best antifouling. In vivo studies on diabetic mice showed accelerated wound healing and skin regeneration with G20S80 modified membranes outperforming chitosan and commercial bandages. These results suggest zwitterionic chitosan soft membranes as advanced wound dressings for diabetic wound management [51].

2.4 Chitosan-based films

The method of solution casting was used to prepare composite films of chitosan and natural modified hydroxyapatite (FeHAp). The SEM, FT-IR, and UV absorption measurements were used to characterize the films. The end result exposed that the films have higher shielding properties as they show excellent UV protection, with ultraviolet protection factor (UPF) values above 50. The antibacterial activity of these films was increased against both Gram-negative and Gram-positive bacteria by FeHAp incorporation; biocompatibility and noncytotoxicity of chitosan were unbothered. The mechanical properties, like flexibility and tensile strength of prepared films, were examined by varying the glycerol and FeHAp amount. The mechanical properties, UV absorption, and porosity of composite films were enhanced by the addition of FeHAp. And all these properties suggested that they could be used as potential wound dressings with improved antibacterial activity [52].

Nanocomposite films of bentonite-chitosan were prepared by using the solution casting method for wound healing applications. Fluid absorption, porosity, and mechanical properties were enhanced due to strong hydrogen bonding between bentonite and chitosan. The physicochemical properties, i.e., water vapor transmission rate (WVTR), folding endurance, and thickness of films, were also examined. The FT-IR was used to study the interaction of chitosan having a positive charge with negatively charged bentonite. Herein, the SEM was used to study the surface morphology of films. The hydrophilic nature of bentonite improved the capabilities of mechanical strength and water absorption of films. The prepared films had improved activity against grams negative and Gram-positive bacteria, which proved their latent use as wound dressing [53].

Chitosan, a biodegradable, antimicrobial, and non-toxic biopolymer, due to its biocompatibility, bioadhesion, and antibacterial properties, is used for wound healing applications. Derived from crustacean waste, it is composed of 2-amino-2-deoxy-D-glucosamine and N-acetyl-2-amino-2-deoxy-D-glucosamine units linked by β-(1,4) glycosidic bonds. Escárcega-Galaz et al. [54] produced chitosan films by solvent casting; they have excellent structural, mechanical, and antimicrobial properties. These films are smooth, transparent, and porous; when combined with agents like glycerol and honey, they are flexible and elongated. Honey adds elasticity and antimicrobial and anti-inflammatory properties; the films are effective against *Pseudomonas aeruginosa* and *Klebsiella pneumoniae*. Chitosan solubility in acidic pH and electrostatic interactions with additives ensure mechanical integrity and optimal fluid retention. The films promote ulcer healing by maintaining moisture balance, controlling gas permeability, and broad-spectrum antimicrobial activity. All these properties, plus low cost and easy production, make chitosan films a promising alternative for advanced wound dressings, especially for skin ulcer treatment [54].

The effects of laser therapy and chitosan-sodium alginate films on burn wound healing in rats were investigated by Dantas et al. [55]. Different natural polymers, when used as wound dressing, improved wound healing. Laser arrays were used to heal the damaged soft tissues. Six groups of male rats were allocated to perform the procedure of burns. Six groups containing untreated, chitosan-alginate-dressed, cellulose dressed (CL), and laser-irradiated combinations were examined over 8 and 14 days. Within 8 days, laser therapy greatly reduced inflammation and stimulated myofibroblastic differentiation. Enhanced blood vessel formation, collagenization, and epithelization were seen in chitosan-alginate films and laser therapy combinations. They also promoted the replacement of type I and type III collagen and superior collagen fiber organization. The wound healing was enhanced by that synergistic approach [55].

2.5 Chitosan-based sponge

Wu et al. [56] developed the CS-CL/HP sponge by merging chitosan (CS), collagen (CL), and heparin (HP), which represented a great improvement in wound healing. Chitosan gives mechanical strength and antibacterial activity, and collagen improves biodegradability and biocompatibility. Heparin then enhanced biological activity by alleviating and shielding growth factors like EGF and VEGF, which are crucial for tissue regeneration. They were synthesized by carbodiimide cross-linking and exhibited broad-spectrum antibacterial activity, improved water absorption, and constant heparin release. In vivo and in vitro studies on guinea pig wound models revealed enhanced growth factor expression, better re-epithelialization, and improved wound healing rates. These results confirmed that CS-CL/HP sponges could be a valuable biomaterial for clinical applications and an effective solution for burn care, as they have improved regenerative capacity, biocompatibility, and antimicrobial activity [56].

Sathiyaseelan et al. [57] synthesized a novel chitosan-based nanocomposite sponge incorporated with phytogenic iron oxide nanoparticles (FeO NPs) from *Pinus densiflora* for wound healing in anemia-associated diabetic patients. These sponges exhibited improved water absorption and high porosity, along with greater antidiabetic, antibacterial, and antioxidant properties and sustained iron release. Enhanced cell proliferation and wound healing potential in HEK-293 cells was confirmed by in vitro studies. The sponges effectively supported oxygen exchange, accelerated tissue regeneration, and reduced microbial infections owing to FeO NPs' bioactivity, polyvinyl alcohol's mechanical strength, and chitosan's antimicrobial activity. These versatile dressings showed potential for diabetic wound care [57].

Cheng et al. [58] combined collagen peptides (COP) and carboxymethyl chitosan (CMC) by chemical coupling and freeze-drying to make advanced burn dressings. These sponges exhibited a porous structure that improved mechanical properties and integration of COP. The CMC prevented infection and scar formation and improved biocompatibility, water solubility, and wound healing. Skin tissue regeneration, proliferation, and cell attachment were promoted by the incorporation of COP. The composite sponges enhanced flexibility and biological activity by addressing individual material limitations. In vitro studies revealed improved cell viability and cell migration. Burn wound experiments were performed in rabbits, and results demonstrated faster wound healing, regenerated epidermis by day 7, and wound recovered completely by day 21. These sponges are cost-effective and hold potential for becoming innovative burn care dressings [58].

2.6 Innovative chitosan formulations for diabetic wounds

Wound healing is delayed in diabetic patients due to an imbalance of systemic and local factors; it leads to diabetic foot ulcer (DFU) in 15–25% of cases. Diabetic foot ulcers (DFU) pose significant challenges for patients' health, as it is the leading cause of non-traumatic amputations worldwide. Chitosan-based wound dressings are proven to be effective treatments because of their distinct properties like biodegradability, biocompatibility, and antimicrobial effects. Chitosan showed significant potential in wound healing by improving tissue regeneration, reducing infection, and controlling inflammation. Innovative chitosan-based composites and hydrogels decreased reactive oxygen species (ROS), demonstrated antibacterial activity against common pathogens like *Escherichia coli* and *Staphylococcus aureus*, and improved re-epithelialization, collagen deposition, and wound closure. Moreover, controlled drug release and improved healing without toxicity were obtained by the integration of other biomaterials and growth factors like polyvinyl alcohol (PVA) in chitosan wound dressings. These developments make chitosan a multifunctional and effective biomaterial for treating DFUs, giving the power to face the complex challenges associated with diabetic wound management [59].

Chijcheapaza-Flores et al. [60] developed injectable chitosan-based hydrogels with Chitosan (CHT) and polymer of β-cyclodextrin (PCD) for cinnamaldehyde's local delivery to treat diabetic foot ulcers (DFU). Significant challenges like low tissue regeneration and infections are associated with the DFU. The cinnamaldehyde (CN), due to its antibacterial activity against Gram-negative and Gram-positive bacteria, was incorporated into these hydrogels to offer anti-biofilm and antimicrobial properties. Notable antimicrobial activity was demonstrated by these hydrogels as bacterial load was reduced by 99.99% when exposed to *Pseudomonas aeruginosa* and *Staphylococcus aureus* for 24 hours. The CN likely acts as a physical crosslinker between chitosan molecules, improved the gel's elasticity, and, as a result, enhanced the viscoelastic properties of the hydrogels. The combination of CN and CHT improved antimicrobial efficacy and drug release dynamics, becoming a potential solution for managing DFU, as revealed by the studies. The treatment is effective and safe, as the hydrogel was cytocompatible, with over 99% cell viability [60].

Wang et al. [61] developed an innovative treatment for diabetic wounds by combining a flexible electrical stimulation (ES) device with Chitosan-Vaseline® gauze (CVG) that improved healing in diabetic foot ulcers. Factors like impaired angiogenesis, inflammation, and epithelialization contribute to the slow healing of diabetic wounds. Electrical stimulation (ES) of chitosan-based dressings is proposed as a solution to these limitations through enhancing cell migration and facilitating wound healing. The Chitosan-Vaseline® gauze dressings maintain moisture retention and possess antibacterial effects, providing a wound microenvironment suited for healing, an important factor in the healing of diabetic wound-infected environments. Wound healing was accelerated by the use of high-voltage monophasic pulsed current (HVMPC) in this elastic dressing that enhanced angiogenesis that involves endothelial cells' (ECs') migration and proliferation. Moreover, the ES treatment turned on key signaling pathways like PI3K/Akt and ERK1/2 that assisted in tissue regeneration and angiogenesis. The flexible ES-chitosan dressing is considered to improve wound healing in diabetic patients as it facilitated wound epithelialization and considerably inhibited scar formation. These findings highlight how combining chitosan-based materials and electrical stimulation could improve treatments for long-lasting diabetic wounds [61].

2.7 Other chitosan-based hybrid materials used for wound dressing

Chen et al. [62] used the solvent precipitation method to develop a novel superabsorbent polymer, exhibiting improved hemostatic properties—carboxymethyl chitosan grafted poly(acrylic acid) (CMCTS-g-PAA). The polymer acquired a porous structure that significantly improved its swelling capacity and hemostatic activity by leveraging the contraction stress of hydrogels during ethanol-induced precipitation. Comparative studies were done on New Zealand white rabbits that demonstrated superior hemostatic properties of the modified polymer in arresting hemorrhage in the arteria cruralis, ear artery, and spleen of New Zealand white rabbits, overtaking conventional wound dressings and unmodified polymers. Leached residuals and $-NH_3^+$ positive charge of chitosan give them high biocompatibility and hemostatic functionality optimized due to their rough porous surface. All these properties make CMCTS-g-PAA a potential material for innovative hemostatic wound dressings, especially in handling severe arterial bleeding [62].

The role of novel chitosan-cellulose composite materials as potential wound dressings has been emphasized by Harkins et al. [63]. Chitosan (CS) is a biopolymer that is best known for its antimicrobial, hemostatic, and wound healing properties, while cellulose (CEL) helps in providing additional mechanistic stability. The CS-CEL composites were synthesized using a green process with a recyclable ionic liquid and no harsh acids or bases, which would otherwise risk damaging the properties of chitosan. The anti-MRSA and anti-VRE properties of the CS-CEL were confirmed as *methicillin resistant Staphylococcus aureus* (MRSA) and *vancomycin resistant Enterococcus* (VRE) strains were present in the panel of bacteria. They also indicated the potential to absorb blood, similar to that of commercial products for wound care, and lyophilized forms had greater absorption capacity than air-dried types. Also, the presence of nano-objects enabled fibroblast growth and reproduction, which is indicative of their ability to drive tissue regeneration in the course of wound healing. The composites also possessed anti-inflammatory properties by lowering levels of tumor necrosis factor and interleukin-6. These findings suggested that CS-CEL composites are environmentally friendly, non-toxic, and biocompatible materials that can be useful for wound healing purposes [63].

Banana peel powder (BPP) was incorporated into chitosan membranes to improve their properties as wound dressing by Kamel et al. [64]. BPP for its antimicrobial and antioxidant properties at concentrations of 0, 2, 5, and 10 wt% mixed with chitosan. The antimicrobial activity of chitosan membranes was improved by adding BPP, with the highest activity at 10 wt%, especially against *Pseudomonas aeruginosa*, *Escherichia coli*, *Staphylococcus aureus*, and *Candida albicans*. Moreover, it became more suitable for controlled wound moisture management because the presence of BPP reduced the swelling behavior of the chitosan matrix. Successful integration of BPP into the chitosan matrix was confirmed through transmission electron microscopy (TEM), SEM, FT-IR, and dielectric spectroscopy, which enhanced its dielectric and mechanical properties. These chitosan/BPP membranes possessed synergistic effects and effective wound healing applications [64].

3. Conclusions

Chitosan-based materials have shown tremendous potential in developing wound healing products, overcoming the drawbacks of standard dressings, and also providing various additional features. Chitosan aids in wound healing as it is biocompatible, biodegradable, hemostatic, and has antimicrobial activity. Wound healing has four phases, i.e., hemostasis, inflammation, proliferation, and remodeling. Chitosan is effective in all of them, which makes it extremely useful in tissue regeneration and wound repairing with minimal scarring. This chapter explores the novel chitosan-based formulations, such as hydrogels, sponges, membranes, and films, that are designed to meet the specific requirements of wound healing. Hydrogels provide a moist environment and drug delivery applications, whereas membranes and films provide mechanical support and antimicrobial activities. High-porosity and absorbent sponges work well for burns and chronic wounds. More effective antibacterial chitosan formulations were made by including nanotechnology and composite materials in them, by increasing their tensile strength, and by controlling their release properties that are especially effective for diabetic wounds and ulcers.

Clinical and preclinical trials summarized in this chapter prove the efficiency of chitosan-based materials, which in turn help to accelerate the wound healing process. These materials decrease the bacterial infections and oxidative stress while boosting processes like collagen deposition, re-epithelialization, angiogenesis, and fibroblast proliferation. Chitosan matrices have gained additional prominence for their potential of targeted drug delivery owing to the encapsulation of bioactive agents, nanoparticles, and growth factors.

Finally, chitosan is a potential biomedical polymer that could enhance twenty-first-century wound care owing to its effective distinction. If the barriers to construction and translation are crossed, wound care approaches targeting diabetic ulcers and chronic wounds could be revolutionized. Chitosan will continue to play an important role in the evolution of new wound healing products due to its adaptability and future development of its dosage form and technology.

Author details

Isha Murtaza[1], Muhammad Yar[2], Zulfiqar Ali Khan[1*] and Sohail Anjum Shahzad[3]

1 Department of Chemistry, Government College University Faisalabad, Faisalabad, Pakistan

2 Interdisciplinary Research Centre in Biomedical Materials, COMSATS University Islamabad, Lahore Campus, Lahore, Pakistan

3 Department of Chemistry, COMSATS University Islamabad, Abbottabad Campus, Abbottabad, Pakistan

*Address all correspondence to: zulfiqarchem@gmail.com

References

[1] Granick MS, Baetz NW, Labroo P, Milner S, Li WW, Sopko NA. In vivo expansion and regeneration of full-thickness functional skin with an autologous homologous skin construct: Clinical proof of concept for chronic wound healing. International Wound Journal. 2019;**16**(3):841-846. DOI: 10.1111/iwj.13109

[2] Moeini A, Pedram P, Makvandi P, Malinconico M, d'Ayala GG. Wound healing and antimicrobial effect of active secondary metabolites in chitosan-based wound dressings: A review. Carbohydrate Polymers. 2020;**233**:115839. DOI: 10.1016/j.carbpol.2020.115839

[3] Feng P, Luo Y, Ke C, Qiu H, Wang W, Zhu Y, et al. Chitosan-based functional materials for skin wound repair: Mechanisms and applications. Frontiers in Bioengineering and Biotechnology. 2021;**18**(9):650598. DOI: 10.3389/fbioe.2021.650598

[4] Chopra H, Bibi S, Kumar S, Khan MS, Kumar P, Singh I. Preparation and evaluation of chitosan/PVA based hydrogel films loaded with honey for wound healing application. Gels. 2022;**8**(2):111. DOI: 10.3390/gels8020111

[5] Alven S, Aderibigbe BA. Chitosan and cellulose-based hydrogels for wound management. International Journal of Molecular Sciences. 2020;**21**(24):9656. DOI: 10.3390/ijms21249656

[6] Bowler PG, Duerden BI, Armstrong DG. Wound microbiology and associated approaches to wound management. Clinical Microbiology Reviews. 2001;**14**(2):244-269. DOI: 10.1128/CMR.14.2.244-269.2001

[7] Fasiku VO, Omolo CA, Devnarain N, Ibrahim UH, Rambharose S, Faya M, et al. Chitosan-based hydrogel for the dual delivery of antimicrobial agents against bacterial methicillin-resistant *Staphylococcus aureus* biofilm-infected wounds. ACS Omega. 2021;**6**(34):21994-22010. Available from: https://pubs.acs.org/doi/epdf/10.1021/acsomega.1c02547

[8] Barrientos S, Stojadinovic O, Golinko MS, Brem H, Tomic-Canic M. Growth factors and cytokines in wound healing. Wound Repair and Regeneration. 2008;**16**(5):585-601. DOI: 10.1111/j.1524-475X.2008.00410.x

[9] Almadani YH, Vorstenbosch J, Davison PG, Murphy AM. Wound healing: A comprehensive review. Seminars in Plastic Surgery. 2021;**35**(03):141-144. DOI: 10.1055/s-0041-1731791

[10] Farokhi M, Mottaghitalab F, Fatahi Y, Khademhosseini A, Kaplan DL. Overview of silk fibroin use in wound dressings. Trends in Biotechnology. 2018;**36**(9):907-922. DOI: 10.1016/j.tibtech.2018.04.004

[11] Kim HS, Sun X, Lee JH, Kim HW, Fu X, Leong KW. Advanced drug delivery systems and artificial skin grafts for skin wound healing. Advanced Drug Delivery Reviews. 2019;**146**:209-239. DOI: 10.1016/j.addr.2018.12.014

[12] Dong R, Guo B. Smart wound dressings for wound healing. Nano Today. 2021;**41**:101290. DOI: 10.1016/j.nantod.2021.101290

[13] Khil M-S, Cha D-I, Kim H-Y, Kim I-S, Bhattarai N. Electrospun nanofibrous polyurethane membrane as wound dressing. Journal of Biomedical Materials Research Part B: Applied Biomaterials. 2003;**67**(2):675-679. DOI: 10.1002/jbm.b.10058

[14] Khan ZA, Jamil S, Akhtar A, Bashir MM, Yar M. Chitosan based hybrid materials used for wound healing applications—A short review. International Journal of Polymeric Materials and Polymeric Biomaterials. 2020;**69**:419-436. DOI: 10.1080/00914037.2019.1575828

[15] Xue H, Hu L, Xiong Y, Zhu X, Wei C, Cao F, et al. Quaternized chitosan-Matrigel-polyacrylamide hydrogels as wound dressing for wound repair and regeneration. Carbohydrate Polymers. 2019;**226**:115302. DOI: 10.1016/j.carbpol.2019.115302

[16] Hussain Z, Thu HE, Shuid AN, Katas H, Hussain F. Recent advances in polymer-based wound dressings for the treatment of diabetic foot ulcer: An overview of state-of-the-art. Current Drug Targets. 2018;**19**(5):527-550. DOI: 10.2174/1389450118666170704132523

[17] Bakhshayesh ARD, Annabi N, Khalilov R, Akbarzadeh A, Samiei M, Alizadeh E, et al. Recent advances on biomedical applications of scaffolds in wound healing and dermal tissue engineering. Artificial Cells, Nanomedicine, and Biotechnology. 2018;**46**(4):691-705. DOI: 10.1080/21691401.2017.1349778

[18] Dagalakis N, Flink J, Stasikelis P, Burke JF, Yannas IV. Design of an artificial skin. Part III. Control of pore structure. Journal of Biomedical Materials Research. 1980;**14**(4):511-528. DOI: 10.1002/jbm.820140417

[19] Huang Y, Onyeri S, Siewe M, Moshfeghian A, Madihally SV. In vitro characterization of chitosan–gelatin scaffolds for tissue engineering. Biomaterials. 2005;**26**(36):7616-7627. DOI: 10.1016/j.biomaterials.2005.05.036

[20] Islam MM, Shahruzzaman M, Biswas S, Sakib MN, Rashid TU. Chitosan based bioactive materials in tissue engineering applications-a review. Bioactive Materials. 2020;**5**(1):164-183. DOI: 10.1016/j.bioactmat.2020.01.012

[21] Sencadas V, Correia DM, Ribeiro C, Moreira S, Botelho G, Ribelles JG, et al. Physical-chemical properties of cross-linked chitosan electrospun fiber mats. Polymer Testing. 2012;**31**(8):1062-1069. DOI: 10.1016/j.polymertesting.2012.07.010

[22] Gu Z, Xie H, Huang C, Li L, Yu X. Preparation of chitosan/silk fibroin blending membrane fixed with alginate dialdehyde for wound dressing. International Journal of Biological Macromolecules. 2013;**58**:121-126. DOI: 10.1016/j.ijbiomac.2013.03.059

[23] Islam MS, Haque P, Rashid TU, Khan MN, Mallik AK, Khan MN, et al. Core–shell drug carrier from folate conjugated chitosan obtained from prawn shell for targeted doxorubicin delivery. Journal of Materials Science: Materials in Medicine. 2017;**28**:55. DOI: 10.1007/s10856-017-5859-x

[24] Rahman MM, Shahruzzaman M, Islam MS, Khan MN, Haque P. Preparation and properties of biodegradable polymer/nano-hydroxyapatite bioceramic scaffold for spongy bone regeneration. Journal of Polymer Engineering. 2019;**39**(2):134-142. DOI: 10.1515/polyeng-2018-0103

[25] Kim Y, Zharkinbekov Z, Raziyeva K, Tabyldiyeva L, Berikova K, Zhumagul D, et al. Chitosan-based biomaterials for tissue regeneration. Pharmaceutics. 2023;**15**(3):807. DOI: 10.3390/pharmaceutics15030807

[26] Croisier F, Jérôme C. Chitosan-based biomaterials for tissue

engineering. European Polymer Journal. 2013;**49**(4):780-792. DOI: 10.1016/j.eurpolymj.2012.12.009

[27] Liu Y, Xiao Y, Cao Y, Guo Z, Li F, Wang L. Construction of chitosan-based hydrogel incorporated with antimonene nanosheets for rapid capture and elimination of bacteria. Advanced Functional Materials. 2020;**30**(35):2003196. DOI: 10.1002/adfm.202003196

[28] Zhang M, Yang M, Woo MW, Li Y, Han W, Dang X. High-mechanical strength carboxymethyl chitosan-based hydrogel film for antibacterial wound dressing. Carbohydrate Polymers. 2021;**256**:117590. DOI: 10.1016/j.carbpol.2020.117590

[29] Jung HY, Le Thi P, HwangBo KH, Bae JW, Park KD. Tunable and high tissue adhesive properties of injectable chitosan based hydrogels through polymer architecture modulation. Carbohydrate Polymers. 2021;**261**:117810. DOI: 10.1016/j.carbpol.2021.117810

[30] Xia L, Wang S, Jiang Z, Chi J, Yu S, Li H, et al. Hemostatic performance of chitosan-based hydrogel and its study on biodistribution and biodegradability in rats. Carbohydrate Polymers. 2021;**264**:117965. DOI: 10.1016/j.carbpol.2021.117965

[31] Drabczyk A, Kudłacik-Kramarczyk S, Głąb M, Kędzierska M, Jaromin A, Mierzwiński D, et al. Physicochemical investigations of chitosan-based hydrogels containing aloe vera designed for biomedical use. Materials. 2020;**13**(14):3073. DOI: 10.3390/ma13143073

[32] Shamloo A, Aghababaie Z, Afjoul H, Jami M, Bidgoli MR, Vossoughi M, et al. Fabrication and evaluation of chitosan/gelatin/PVA hydrogel incorporating honey for wound healing applications: An in vitro, in vivo study. International Journal of Pharmaceutics. 2021;**592**:120068. DOI: 10.1016/j.ijpharm.2020.120068

[33] Silva AO, Cunha RS, Hotza D, Machado RA. Chitosan as a matrix of nanocomposites: A review on nanostructures, processes, properties, and applications. Carbohydrate Polymers. 2021;**272**:118472. DOI: 10.1016/j.carbpol.2021.118472

[34] Sakthiguru N, Sithique MA. Preparation and in vitro biological evaluation of Lawsone loaded O-carboxymethyl chitosan/zinc oxide nanocomposite for wound-healing application. ChemistrySelect. 2020;**5**(9):2710-2718. DOI: 10.1002/slct.201904159

[35] Shah A, Buabeid MA, Arafa ES, Hussain I, Li L, Murtaza G. The wound healing and antibacterial potential of triple-component nanocomposite (chitosan-silver-sericin) films loaded with moxifloxacin. International Journal of Pharmaceutics. 2019;**564**:22-38. DOI: 10.1016/j.ijpharm.2019.04.046

[36] Nowroozi N, Faraji S, Nouralishahi A, Shahrousvand M. Biological and structural properties of graphene oxide/curcumin nanocomposite incorporated chitosan as a scaffold for wound healing application. Life Sciences. 2021;**264**:118640. DOI: 10.1016/j.lfs.2020.118640

[37] Azadmanesh F, Pourmadadi M, Zavar Reza J, Yazdian F, Omidi M, Haghirosadat BF. Synthesis of a novel nanocomposite containing chitosan as a three-dimensional printed wound dressing technique: Emphasis on gene expression. Biotechnology Progress. 2021;**37**(4):e3132. DOI: 10.1002/btpr.3132

[38] Kędzierska M, Blilid S, Miłowska K, Kołodziejczyk-Czepas J, Katir N, Lahcini M, et al. Insight into factors influencing wound healing using phosphorylated cellulose-filled-chitosan nanocomposite films. International Journal of Molecular Sciences. 2021;**22**(21):11386. DOI: 10.3390/ijms222111386

[39] Mohandas A, Deepthi S, Biswas R, Jayakumar R. Chitosan based metallic nanocomposite scaffolds as antimicrobial wound dressings. Bioactive Materials. 2018;**3**(3):267-277. DOI: 10.1016/j.bioactmat.2017.11.003

[40] Muzzarelli RA, Morganti P, Morganti G, Palombo P, Palombo M, Biagini G, et al. Chitin nanofibrils/chitosan glycolate composites as wound medicaments. Carbohydrate Polymers. 2007;**70**(3):274-284. DOI: 10.1016/j.carbpol.2007.04.008

[41] Rahimi M, Ahmadi R, Kafil HS, Shafiei-Irannejad V. A novel bioactive quaternized chitosan and its silver-containing nanocomposites as a potent antimicrobial wound dressing: Structural and biological properties. Materials Science and Engineering: C. 2019;**101**:360-369. DOI: 10.1016/j.msec.2019.03.092

[42] Dutta J, Devi N. Preparation, optimization, and characterization of chitosan-sepiolite nanocomposite films for wound healing. International Journal of Biological Macromolecules. 2021;**186**:244-254. DOI: 10.1016/j.ijbiomac.2021.07.020

[43] Sandri G, Aguzzi C, Rossi S, Bonferoni MC, Bruni G, Boselli C, et al. Halloysite and chitosan oligosaccharide nanocomposite for wound healing. Acta Biomaterialia. 2017;**57**:216-224. DOI: 10.1016/j.actbio.2017.05.032

[44] Shanmugapriya K, Kim H, Saravana PS, Chun BS, Kang HW. Fabrication of multifunctional chitosan-based nanocomposite film with rapid healing and antibacterial effect for wound management. International Journal of Biological Macromolecules. 2018;**118**:1713-1725. DOI: 10.1016/j.ijbiomac.2018.07.018

[45] Vedakumari WS, Prabu P, Sastry TP. Chitosan-fibrin nanocomposites as drug delivering and wound healing materials. Journal of Biomedical Nanotechnology. 2015;**11**(4):657-667. DOI: 10.1166/jbn.2015.1948

[46] Zhai M, Xu Y, Zhou B, Jing W. Keratin-chitosan/n-ZnO nanocomposite hydrogel for antimicrobial treatment of burn wound healing: Characterization and biomedical application. Journal of Photochemistry and Photobiology, B: Biology. 2018;**180**:253-258. DOI: 10.1016/j.jphotobiol.2018.02.018

[47] Vedula SS, Yadav GD. Chitosan-based membranes preparation and applications: Challenges and opportunities. Journal of the Indian Chemical Society. 2021;**98**(2):100017. DOI: 10.1016/j.jics.2021.100017

[48] Augustine R, Rehman SR, Ahmed R, Zahid AA, Sharifi M, Falahati M, et al. Electrospun chitosan membranes containing bioactive and therapeutic agents for enhanced wound healing. International Journal of Biological Macromolecules. 2020;**156**:153-170. DOI: 10.1016/j.ijbiomac.2020.03.207

[49] Behera SS, Das U, Kumar A, Bissoyi A, Singh AK. Chitosan/TiO_2 composite membrane improves proliferation and survival of L929 fibroblast cells: Application in wound dressing and skin regeneration. International Journal of Biological

Macromolecules. 2017;**98**:329-340. DOI: 10.1016/j.ijbiomac.2017.02.017

[50] Meng X, Tian F, Yang J, He CN, Xing N, Li F. Chitosan and alginate polyelectrolyte complex membranes and their properties for wound dressing application. Journal of Materials Science: Materials in Medicine. 2010;**21**:1751-1759. DOI: 10.1007/s10856-010-3996-6

[51] Lin H-T, Venault A, Chang Y. Zwitterionized chitosan based wound dressings for diabetic wound healing. Journal of Membrane Science. 2019;**591**:117319. DOI: 10.1016/j.memsci.2019.117319

[52] Cunha CS, Castro PJ, Sousa SC, Pullar RC, Tobaldi DM, Piccirillo C, et al. Films of chitosan and natural modified hydroxyapatite as effective UV-protecting, biocompatible and antibacterial wound dressings. International Journal of Biological Macromolecules. 2020;**159**:1177-1185. DOI: 10.1016/j.ijbiomac.2020.05.077

[53] Devi N, Dutta J. Preparation and characterization of chitosan-bentonite nanocomposite films for wound healing application. International Journal of Biological Macromolecules. 2017;**104**:1897-1904. DOI: 10.1016/j.ijbiomac.2017.02.080

[54] Escárcega-Galaz AA, Sánchez-Machado DI, López-Cervantes J, Sanches-Silva A, Madera-Santana TJ, Paseiro-Losada P. Mechanical, structural and physical aspects of chitosan-based films as antimicrobial dressings. International Journal of Biological Macromolecules. 2018;**116**:472-481. DOI: 10.1016/j.ijbiomac.2018.04.149

[55] Dantas MDM, Cavalcante DRR, Araújo FEN, Barretto SR, Aciole GTS, Pinheiro ALB, et al. Improvement of dermal burn healing by combining sodium alginate/chitosan-based films and low level laser therapy. Journal of Photochemistry and Photobiology, B: Biology. 2011;**105**(1):51-59. DOI: 10.1016/j.jphotobiol.2011.06.009

[56] Wu K, Zhao D, Cui H. Preparation and evaluation of heparinized sponge based on collagen and chitosan for wound healing. Journal of Bioactive and Compatible Polymers. 2020;**35**(4-5):314-327. DOI: 10.1177/08839115209399

[57] Sathiyaseelan A,Saravanakumar K, Mariadoss AVA, Wang M-H. Antimicrobial and wound healing properties of FeO fabricated chitosan/PVA nanocomposite sponge. Antibiotics. 2021;**10**(5):524. DOI: 10.3390/antibiotics10050524

[58] Cheng Y, Hu Z, Zhao Y, Zou Z, Lu S, Zhang B, et al. Sponges of carboxymethyl chitosan grafted with collagen peptides for wound healing. International Journal of Molecular Sciences. 2019;**20**(16):3890. DOI: 10.3390/ijms20163890

[59] Silva JD, Leal EC, Carvalho E, Silva EA. Innovative functional biomaterials as therapeutic wound dressings for chronic diabetic foot ulcers. International Journal of Molecular Sciences. 2023;**24**(12):9900. DOI: 10.3390/ijms24129900

[60] Chijcheapaza-Flores H, Tabary N, Chai F, Maton M, Staelens JN, Cazaux F, et al. Injectable chitosan-based hydrogels for trans-cinnamaldehyde delivery in the treatment of diabetic foot ulcer infections. Gels. 2023;**9**(3):262. DOI: 10.3390/gels9030262

[61] Wang X-F, Li M-L, Fang Q-Q, Zhao W-Y, Lou D, Hu Y-Y, et al. Flexible electrical stimulation device with chitosan-Vaseline® dressing accelerates wound healing in diabetes.

Bioactive Materials. 2021;**6**(1):230-243. DOI: 10.1016/j.bioactmat.2020.08.003

[62] Chen Y, Zhang Y, Wang F, Meng W, Yang X, Li P, et al. Preparation of porous carboxymethyl chitosan grafted poly (acrylic acid) superabsorbent by solvent precipitation and its application as a hemostatic wound dressing. Materials Science and Engineering: C. 2016;**63**:18-29. DOI: 10.1016/j.msec.2016.02.048

[63] Harkins AL, Duri S, Kloth LC, Tran CD. Chitosan–cellulose composite for wound dressing material. Part 2. Antimicrobial activity, blood absorption ability, and biocompatibility. Journal of Biomedical Materials Research Part B: Applied Biomaterials. 2014;**102**(6):1199-1206. DOI: 10.1002/jbm.b.33103

[64] Kamel NA, Abd El-messieh SL, Saleh NM. Chitosan/banana peel powder nanocomposites for wound dressing application: Preparation and characterization. Materials Science and Engineering: C. 2017;**72**:543-550. DOI: 10.1016/j.msec.2016.11.104

Chapter 3

Chitosan-Pectin Composite Biomaterials: Types, Interactions, Formulations, and Biomedical Applications

Lăcrămioara Popa, Mihaela Violeta Ghica, Cristina-Elena Dinu-Pîrvu and Irina-Alexandra Dumitrescu

Abstract

Chitosan and pectin are two of the most abundant polysaccharides available in the natural world. Chitosan is readily obtained from crustacean shells, while pectin has a fundamental and supportive function in the plant cell wall and is extracted with satisfactory yields from apple pomace and citrus peel. Recent focus on natural polymers for their ecologically friendly properties (biocompatibility and biodegradability) brought forward in the pharmaceutical and medical fields a wide range of new materials and combinations, that had previously been overlooked by researchers in favor of synthetic polymers. Natural polymers present challenges due to their complex structures and variability but also have great versatility and can be tailored to suit particular needs. Chitosan and pectin have been comprehensively investigated independently, with their characteristics thoroughly examined. They have now been combined into a composite to determine how their advantageous properties can be mutually enhanced. This chapter aims to review chitosan-pectin composites emphasizing the interaction mechanisms between the two polymers, exploring formulation strategies and biomedical/biotechnological applications.

Keywords: chitosan, pectin, biomaterials, composites, biomedical applications

1. Introduction

Chitosan-pectin (Ch-Pec) composites have emerged as versatile biomaterials by combining two abundant natural polysaccharides-pectin (from plant cell walls, e.g., citrus or apple peels) and chitosan (from chitin in crustacean shells) [1]. Both polymers are nontoxic, biocompatible, biodegradable and ecologically friendly [2, 3]. After years of separate research, recent studies have explored the potential of combining these materials.

Since both are approved as safe for use in food and pharmaceutical products, chitosan-pectin composites provide a wide range of purposes [4, 5]. By blending these

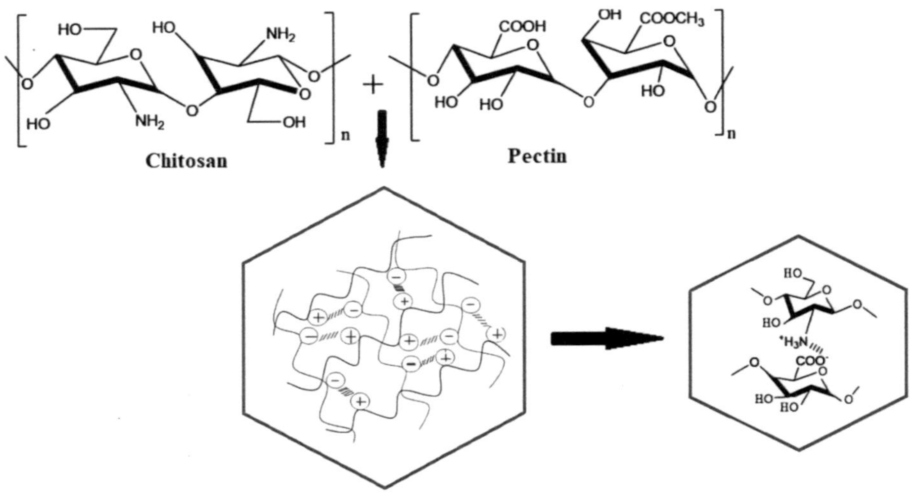

Figure 1.
The graphical description of the interaction mechanisms that underlie Ch-Pec network formation, adapted from Soubhagya et al. [9].

oppositely charged polymers, researchers have created hybrid materials in various formats – from injectable hydrogels [6, 7] and porous scaffolds [8] to thin films/membranes and nano/microparticles [9, 10] – for a wide range of biomedical and biotechnological applications, that include drug delivery systems, tissue engineering scaffolds, wound dressings [11], encapsulation of bioactive [12], and antimicrobial coatings [13], among others.

Recent years have witnessed significant research focused on optimizing pectin-chitosan formulations and elucidating their behavior. This chapter provides a comprehensive overview of the past decade's advances, covering: (1) the types of pectin used (high- vs. low-methoxy, amidated, etc.) and their roles in combination with chitosan; (2) the interaction mechanisms (e.g., ionic crosslinking and polyelectrolyte complexation) that underlie Ch-Pec network formation, as shown in **Figure 1**; (3) key formulation strategies such as hydrogels, nanoparticles, scaffolds, coatings, and films; (4) major application areas (drug delivery, wound healing, tissue engineering, encapsulation of bioactives, and antimicrobial materials), with highlights from recent studies; (5) a discussion of advantages and challenges of Ch-Pec systems in comparison to other biopolymer combinations.

Pectin-chitosan composites are a promising and highly tunable class of biomaterials that utilize the complementary properties of a negatively charged plant polysaccharide and a positively charged marine polysaccharide for advanced therapeutic and biotechnological applications [5].

2. Types of pectin used and their roles with chitosan

Pectin is a complex *anionic* polysaccharide rich in galacturonic acid units (some methyl-esterified) [1, 14]. Pectin is categorized by degree of methyl esterification (DM): high-methoxy pectin (HMP) has DM >50%, whereas low-methoxy pectin (LMP) has DM <50% [1]. Additionally, amidated pectin is a modified LMP in which some galacturonic acid carboxyl groups are converted to amides [15]. These structural differences strongly influence how pectin interacts with chitosan and their function in composites:

High-methoxyl pectin (HMP): With many esterified carboxyl, HMP carries fewer negative charges. It typically gels in acidic, high-sugar conditions *via* hydrogen bonding and hydrophobic interactions [1]. In Ch-Pec systems, HMP can still form polyelectrolyte complexes with chitosan if some carboxyl remains deprotonated. Notably, HMP has been used for colon-targeted drug delivery: one study showed HMP-based coatings produced a delayed drug release specifically in the colon with no premature release in the upper GI tract [16, 17]. Thus, HMP's lower solubility in gastric fluids can be advantageous in protecting drugs until reaching higher pH environments.

Low-methoxyl pectin (LMP): LMP provides more free carboxyl groups ($-COO^-$) for ionic interaction. It readily forms ionotropic gels with multivalent cations like Ca^{2+} (egg-box crosslinking) and can be obtained within a wider pH range, from 3.5 to 9.5 [1, 18]. In combination with Chitosan, LMP's abundant negative sites strongly bind chitosan's protonated amines, yielding robust polyelectrolyte complexes (PECs). LMP is commonly used in Ch-Pec hydrogels, beads, and nanoparticles to maximize electrostatic crosslinking and pH-responsive behavior. For example, pectin with low esterification has been incorporated to achieve pH-sensitive drug release in the colon (enzymatic degradation by colonic pectinases) [19]. LMP contributes to firmer gels and slower dissolution in neutral/basic conditions, making it ideal for sustained-release matrices [20].

Amidated pectin (amidated LMP): Amidated pectin (often denoted LMA) has a portion of carboxyl converted to amid groups, slightly reducing the charge density but conferring unique gelling properties (e.g., requiring less Ca^{2+}). Amidated pectin remains water-soluble and can still interact electrostatically with chitosan. Studies report that amidated low-methoxy pectin yields nanoparticles with optimal size and stability for drug delivery [21]. In multilayer Ch-Pec films, using amidated pectin improved uniformity and mechanical strength [20]. Thus, amidated pectin can enhance processability of Ch-Pec systems, forming smoother complexes and potentially modulating the *rate* of Ch-Pec interactions (due to slightly fewer charge sites).

In all cases, source and structure of pectin play an important role. Plant pectin contains "smooth" homogalacturonan regions and "hairy" rhamnogalacturonan regions with neutral sugar side chains [1, 22]. These structural nuances can affect composite properties. For instance, highly branched pectin may yield softer, more flexible gels, whereas long homogalacturonan chains favor tighter crosslinking with chitosan. Pectin's *molecular weight* and *purity* are also important – modern standardized pectin grades with tailored DM and molecular weight are now available, mitigating the batch-to-batch variability that historically challenged reproducibility [19]. Overall, researchers choose the type of pectin based on the desired outcome: HMP for certain targeted release profiles, LMP for strong ionic networks, and amidated pectin for improved handling and film properties. In combination with chitosan (which is cationic and itself varies by molecular weight and degree of deacetylation), these pectin types allow fine-tuning of composite behavior.

3. Interaction mechanisms between pectin and chitosan

The primary mechanism driving Ch-Pec composite formation is electrostatic interaction between the oppositely charged functional groups of the polymers. Chitosan is a polycationic polysaccharide – in acidic to mid-range pH it carries protonated amine groups (–NH^{3+}) along its chain [23]. Pectin is polyanionic – above a certain pH, its galacturonic acid residues ionize to carboxylate anions (–COO–). When mixed under appropriate conditions, ionic crosslinks spontaneously form between –NH^{3+} of chitosan and –COO– of pectin, leading to a polyelectrolyte complex (PEC) network [1].

pH-Dependence: The Ch-Pec electrostatic bond formation is highly pH-dependent. Pectin's carboxyl groups have a pKa ≈ 3.5–3.6, so at pH above ~3.6, a significant fraction is deprotonated (–COO^-) and capable of ionic bonding. Chitosan's amine groups have pKa ~6.3–6.5; at pH below that, many are protonated (–NH3+) [24, 25]. Optimal complexation occurs in moderately acidic pH (roughly 3 < pH < 6) where pectin carries negative charges and chitosan carries positive charges. If the pH is too low (<<3.5), pectin becomes mostly neutral (–COOH) and cannot interact, while chitosan remains charged. If pH is too high (> ~6.5), chitosan loses its charge (–NH_2), so the polymers no longer attract. This pH-sensitive ionic gelation gives Ch-Pec systems their environment-responsive behavior: for example, a Ch-Pec hydrogel may be stable in an acidic stomach (where the complex stays intact) but swell or dissociate in neutral intestinal conditions (releasing its load) – a mechanism exploited for colon-targeted delivery [19].

Polyelectrolyte complexation and coacervation occur when chitosan and pectin are mixed in solution, leading to immediate charge attraction that may result in the coacervation or precipitation of insoluble polyelectrolyte complex particles if not properly managed [20]. Researchers achieve this through controlled mixing techniques, such as the dropwise addition of one polymer to another or the use of dilute concentrations, to produce stable nanoparticles or homogeneous hydrogels instead of macroscopic precipitates [20]. Soluble PECs, characterized as non-stoichiometric complexes, may form when one polymer is present in excess, resulting in a stable colloidal system. Pectin-rich or chitosan-rich complexes remain dispersed due to their overall charge, which inhibits aggregation [26]. A study demonstrated that altering the Ch:Pec mass ratio from 3:1 to 1:3 influenced nanoparticle size (30–330 nm) and zeta potential (–29 mV to +73 mV); a 1:1 ratio was found to be the most effective carrier for a chemotherapeutic drug. This highlights the impact of stoichiometry on assembly: at a 1:1 ratio, the neutralized PEC may aggregate into a compact nanoparticle, optimizing drug loading and stability [27].

Ionic gelation and crosslinking agents: The Ch-Pec network can also be reinforced by multivalent ions. Low-Methoxyl pectin domains crosslink *via* Ca^{2+} ions (egg-box model), so in a composite gel, Ca^{2+} can further tighten the pectin sections. Similarly, chitosan is often ionically crosslinked with tripolyphosphate (TPP) in nanoparticle formulations; a Ch-Pec-TPP system can have both polymer-polymer and polymer-ion links. However, a key advantage of Ch-Pec PECs is that they often gel without any chemical crosslinkers, avoiding potential toxicity [20]. The ionic gelation is mild and aqueous, ideal for sensitive drugs or cells. For example, pectin-chitosan hydrogel beads have been made by dropping pectin into chitosan (or vice versa) solution – an interfacial coacervation process that immediately forms a gelled bead coated by the second polymer [19].

Secondary interactions: Besides ionic bonds, hydrogen bonding and hydrophobic interactions also contribute to Ch-Pec network stability [1]. Pectin's backbone and chitosan's hydroxyl groups can form hydrogen bonds, while any methyl-esterified or acetylated regions may add hydrophobic contacts. These secondary forces improve the integrity of the complex, especially under low-ion conditions. Indeed, analyses by FTIR and rheology confirm that ionic bonds are the primary crosslinks, but additional intermolecular forces exist in hydrated Ch-Pec gels [26].

Covalent crosslinking (chemical enhancements): Chemical crosslinkers are selectively used in some formulations to covalently link pectin and chitosan, resulting in a semi-interpenetrating or dual-network hydrogel. Carbodiimide coupling (EDC/NHS) facilitates the formation of amide bonds between the –COOH groups of pectin and the $-NH_2$ groups of chitosan [1]. Oxidized pectin (containing aldehyde groups) can interact with the amines of chitosan to establish Schiff-base links (imine bonds) [28]. One study created an *in situ* gelling injectable hydrogel using oxidized chitosan and amidated pectin, which crosslinked *via* imine bonds and showed self-healing properties [7]. Such covalent linking is used to strengthen the hydrogel (preventing dissolution even when ionic bonds are screened by high salt or pH changes). Despite this, the first self-assembly is often driven by ionic affinity to bring the polymers into close proximity even using covalent techniques. In Ch-Pec systems, ionic gelation/polyelectrolyte complexation is the fundamental process; usually enhanced by secondary bonding or covalent linkages to suit the mechanical and release characteristics.

4. Formulation strategies for pectin: Chitosan systems

Thanks to their complementary chemistry, pectin and chitosan can be co-assembled into a variety of formulations. Key formulation strategies include hydrogels (bulk or injectable), nanoparticles and microparticles, beads, films/coatings, multilayer assemblies, nanofibers, and porous scaffolds. Researchers select a formulation based on the intended application (e.g., an oral drug nanoparticle vs. a wound dressing film). Below we outline common formulation types and methods (**Figure 2**).

Hydrogels (Bulk and Injectable): Ch-Pec hydrogels are three-dimensional networks capable of holding large amounts of water. They can be formed by simply mixing solutions of chitosan and pectin in the proper pH range (usually in dilute acid). The result is a physical hydrogel where the polymers are crosslinked by ionic bonds (a physical gel with no external crosslinker) [1]. Hydrogels can be cast in molds to create bulk gel disks or slabs for use as scaffolds. Alternatively, *injectable* hydrogels are

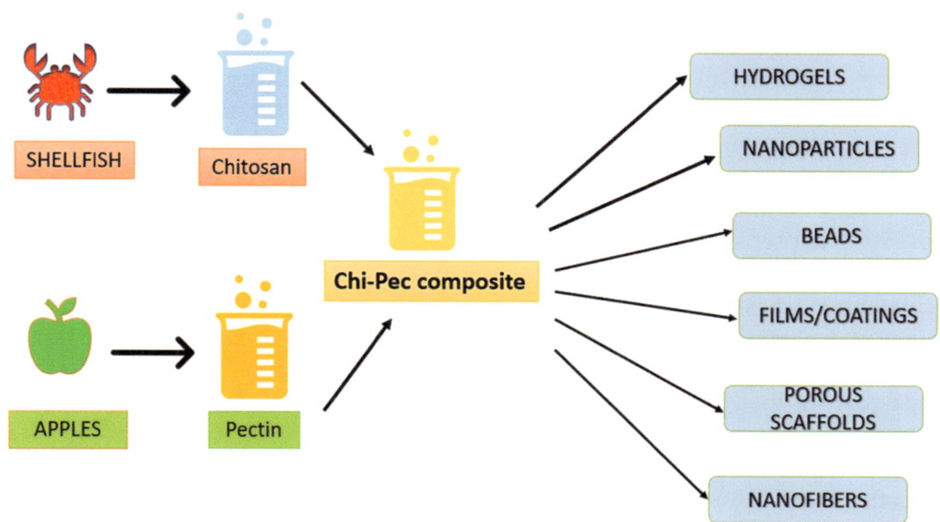

Figure 2.
Common Ch-Pec composites formulation types.

formulated to be liquid during handling and gel upon a trigger (often pH or temperature change) at the target site. For instance, thermosensitive Ch-Pec hydrogels have been developed that are injectable at room temperature and solidify at body temperature. One approach achieved this by partially neutralizing chitosan with a weak base so that the mixture remains liquid when cold but gels upon warming [6]. These injectable Ch-Pec hydrogels have been studied for minimally invasive delivery of cells and drugs in tissue engineering [27]. In general, hydrogel properties (stiffness, swelling, degradation) are tuned by the Ch:Pec ratio, polymer concentration, and any added crosslinkers.

Nanoparticles and microparticles: Utilizing *ionic self-assembly*, researchers have developed pectin-chitosan nanoparticles (NPs) for the purpose of drug delivery. A common method is ionotropic gelation: for example, adding a pectin solution to chitosan under stirring leads to the formation of sub-micron PEC particles. The size and charge of these NPs depend on mixing conditions and polymer ratios [26]. *Sonication* or *high-shear mixing* can further reduce particle size. These NPs can be prepared in the presence of a drug so that the drug is encapsulated during complexation (often achieving high loading for oppositely charged drugs *via* electrostatic attraction). Spray-drying is another technique: a blended Ch-Pec solution can be atomized and dried to yield composite microparticles. For instance, spray-dried chitosan-pectin microspheres have been explored for controlled release of theophylline [28]. Ch-Pec nanoparticles have been reported in sizes from ~50 nm up to a few microns, with surface charge tunable between positive or negative by varying polymer ratios [26, 29] Layer-by-layer (LbL) coating can create core-shell particles: for example, chitosan NP cores coated with pectin shell or vice versa. Pachaleva et al. designed *nisin-loaded* pectin NPs and then coated them with chito-oligosaccharide, producing stable particles with either positive or negative surface charge depending on coating thickness. Such multilayer NPs are useful to modulate mucoadhesion and drug release kinetics.

Hydrogel beads: These are millimeter-scale gel particles often made by extrusion or dripping. For example, dropping a solution of low-methoxy pectin containing a drug into a chitosan solution will form beads with a pectin core and chitosan coating (or vice versa). *Electrostatic or electrospray dripping* can produce smaller, more uniform beads by applying a voltage to the droplet formation [19]. Beads combine the benefits of hydrogels (high water content) and particulate carriers (discrete dosage forms). Chitosan-coated pectin beads have been studied for oral delivery of probiotics and drugs: the chitosan shell delays release in acidic stomach and then the bead disintegrates in the intestine [19]. These beads are typically stabilized by ionic crosslinks (e.g., the pectin core crosslinked with Ca^{2+} plus the polyelectrolyte coating. Sizes range from ~1 mm (with electrodripping) up to several millimeters. For instance, Acevedo et al. prepared 3–5 mm pectin-chitosan hydrogel beads by interfacial complexation; the beads remained intact in simulated gastric fluid and only released their payload upon transitioning to intestinal conditions [19].

Films, membranes, and coatings: By casting or depositing Ch-Pec mixtures, one can form thin films or membranes. These can serve as wound dressing films, buccal or transdermal patches, food packaging, or implant coatings. Solvent casting is common: pectin and chitosan are co-dissolved (or layered sequentially) and the solvent evaporated to leave a composite film. To improve uniformity, layer-by-layer (LbL) assembly is often used: alternating layers of pectin and chitosan are deposited to build up a stratified film with well-controlled architecture. Such polyelectrolyte multilayer films ensure homogeneous mixing at the nanoscale and avoid bulk precipitation issues. An example is *chitosan/low-methoxy amidated pectin multilayer films* optimized for buccal drug delivery – these films had higher mechanical strength and thermal stability than single-polymer films [20]. Electro spraying is another method to coat surfaces: Alsharbaty et al. used electrospray deposition to coat titanium implant surfaces with a pectin-chitosan composite layer. The resulting *PCPC coating* was homogeneous, strongly adherent, and endowed the surface with antibacterial properties [4]. In general, Ch-Pec films are flexible, transparent, and bio adhesive, making them suitable for covering wounds or mucosal tissues.

Chitosan often improves the tensile strength and water-resistance of pectin films, while pectin can enhance the film's stability in acidic environments by preventing chitosan from dissolving. For instance, adding chitosan to pectin films significantly increased their mechanical strength and reduced water vapor [30].

Porous Scaffolds (Foams): By freeze-drying or gas-foaming Ch-Pec hydrogels, lightweight sponges or foams can be created. These have high porosity and surface area, ideal for tissue engineering (allowing cell seeding and nutrient flow) or as absorbent wound dressings. Freeze-drying (lyophilization) of a Ch-Pec gel yields a scaffold with a network of pores whose size can be tuned by freezing rate. A 2011 study prepared freeze-dried pectin-chitosan scaffolds and showed they supported attachment and proliferation of human osteoblasts, indicating potential in bone tissue engineering [1]. Similarly, highly porous chitosan/pectin sponges were developed for wound healing – one report notes a 70:30 chitosan: pectin sponge had suitable porosity and successfully healed wounds in an animal model [31]. The combination of chitosan's hemostatic property and pectin's gel-forming ability makes such sponges attractive for bleeding wounds. Porous Ch-Pec scaffolds can also be reinforced with additives (e.g., bio glass for bone, or nanocellulose for strength to enhance mechanical or biological performance [32].

Nanofibers: Although less common than hydrogels or films, electrospinning can produce ultrafine Ch-Pec fibers for mats or meshes. Pure chitosan is difficult to electro spin, but blending with pectin (and often a co-solvent or carrier polymer) can enable fiber formation [33]. Electrospun pectin-chitosan mats have potential as wound dressings (high surface area for absorption and cell interaction) or as filtration membranes with antimicrobial activity. Martău et al. noted that combining chitosan, alginate, and pectin in electrospun nanofibers yields biomaterials with excellent biocompatibility and bioadhesion [34]. This area is still emerging, but it offers a way to create nanostructured composite surfaces.

The most commonly developed formulation types and preparation methods are summarized in **Table 1**.

Formulation type	Preparation method	Application area	Ref.
Hydrogels	Ionic gelation, pH-triggered mixing	Wound dressing, drug delivery	[1, 6, 27]
Nanoparticles	Ionotropic self-assembly, spray-drying	Colon targeting, anticancer therapy	[26, 28, 29]
Beads	Extrusion dripping, electrodripping	Probiotic/nutraceutical delivery	[19]
Films/Coatings	Solvent casting, LbL deposition	Transdermal patches, implant coatings	[4, 20, 30, 35–37]
Porous Scaffolds	Freeze-drying hydrogels	Tissue scaffolds (bone, skin)	[1, 31, 32, 38, 39]
Nanofibers	Electrospinning with carriers	Wound mats, filtration membranes	[33, 34]

Table 1.
Formulation types and preparation methods.

5. Biomedical and biotechnological applications

Pectin-chitosan composites have been explored across numerous pharmaceutical, biomedical, and biotechnological applications. Their biocompatibility, mucoadhesiveness, tunable gelation, and ease of functionalization make them suitable for delivering therapeutics, supporting tissue regeneration, protecting bioactive, and preventing infections. Below, we review major application areas, highlighting how Ch-Pec systems are utilized in each:

5.1 Drug delivery systems

One of the most extensively studied uses of Ch-Pec composites is as drug delivery systems. The Ch-Pec combination offers several advantages: pH-responsive swelling/degradation, enzymatic degradability (pectin by pectinases, chitosan by lysozyme), and mucoadhesiveness (especially from chitosan). These properties have been harnessed to create targeted and controlled-release drug carriers:

Oral drug delivery and colon targeting: Because of its favorable characteristics—biocompatibility, mucoadhesiveness, adjustable gelation, and simplicity of functionalization—pectin-chitosan composites have been intensively investigated in biotechnology, medicine, and pharmacology. These properties fit them for treatment of diseases, tissue regeneration, bioactive chemical protection, and infection prevention. Ch-Pec composites find one of the most investigated applications in drug delivery systems. Designing regulated and targeted drug release systems calls for their pH-responsive swelling and disintegration, enzymatic breakdown (pectin by pectinase, chitosan by lysozyme), and strong mucoadhesive behavior (particularly from chitosan).

Particularly in oral medication administration and colon targeting, pectin is quite fit for colon-specific drug delivery as it resists digestion in the upper gastrointestinal tract and is fermented by colonic bacteria.

Controlled release and mucoadhesive delivery: The interaction between chitosan and pectin can markedly reduce drug diffusion, facilitating sustained release profiles. Ch-Pec hydrogels and films have been loaded with various drugs, including small molecules and proteins, exhibiting prolonged release characteristics. Resveratrol-loaded chitosan-pectin core-shell nanoparticles demonstrated a sustained release of resveratrol for approximately 30 hours *in vitro*, significantly exceeding the release duration observed from non-complexed matrices [40]. The mucoadhesive properties of chitosan enhance retention at absorption sites, such as the buccal and intestinal mucosa, thereby improving bioavailability. Kaur et al. developed a mucoadhesive buccal film utilizing an interpolymer complex of chitosan and pectin for the delivery of carvedilol, achieving a bioavailability increase of approximately 2.14 times relative to the oral route [41]. Ch-Pec nanoparticles have been utilized for nasal and ocular drug delivery, where mucoadhesion plays a critical role. An *ex vivo* study demonstrated that thiolated pectin-chitosan nanoparticles for ocular delivery exhibited improved corneal permeation of a drug compared to a simple solution [42].

Targeted Cancer Therapy: Interest in the application of pectin-chitosan carriers for cancer drug delivery is increasing. Modifying surface charge and attaching targeting ligands renders Ch-Pec nanoparticles an effective platform. Silant'ev et al. developed chitosan-pectin nanoparticles for the targeted delivery of temozolomide (TMZ) to brain tumors. Through the optimization of the polymer ratio, approximately 200 nm cationic particles were produced, capable of traversing biological barriers and facilitating controlled release of TMZ in simulated body fluids. The particles enhanced drug immobilization and release kinetics, indicating a promising approach for brain cancer, particularly in overcoming the challenges of targeted delivery across the blood-brain barrier [27]. Researchers employed an alternative method by encapsulating curcumin, a hydrophobic anticancer compound, within pectin-chitosan nanogels. This approach resulted in improved uptake and cytotoxicity in MCF-7 breast cancer cells when compared to free curcumin [43, 44]. The positive charge of chitosan aids in the binding of nanoparticles to negatively charged cell membranes, thereby facilitating endocytosis. Additionally, pectin can be modified with targeting moieties, such as folate, to direct nanoparticles specifically to cancer cells.

Multi-component delivery systems: Pectin-chitosan systems are utilized for the delivery of complex components, including vaccines, peptides, and living cells. Macrophage-targeted liposome-loaded Ch-Pec microgels have been developed for oral delivery in colitis treatment. In this system, pectin-chitosan functions as a protective gel matrix for liposomes, facilitating their release in the colon for uptake by inflammatory cells [45].

Pectin-chitosan drug delivery systems provide biocompatible and tunable vehicles capable of responding to pH or enzymes, adhering to mucosal surfaces, and protecting drugs from premature release or degradation. The variety of formats—hydrogels, films, beads, nanoparticles—facilitates customization for particular routes of administration and therapeutic requirements. Numerous systems have shown enhanced pharmacokinetics, efficacy, or diminished side effects in preclinical models relative to conventional dosage forms, highlighting the significance of Ch-Pec composites in advanced drug delivery.

Wound healing applications. Wound healing necessitates a precise equilibrium of moisture, infection prevention, and facilitation of tissue regeneration. Pectin-chitosan composites meet several essential criteria for an optimal wound dressing: they are biodegradable, capable of creating a moist gel environment, nontoxic, and can possess antimicrobial and bioactive release characteristics. Various types of Ch-Pec dressings have been investigated.

Hydrogel dressings. Hydrogels effectively maintain moisture in wounds, facilitating re-epithelialization, and adapt to irregular wound contours. Ch-Pec hydrogels possess an inherent water content exceeding 95% and demonstrate the capacity to absorb excess wound exudate, with one formulation capable of absorbing approximately 150 times its weight in fluid [28]. Chitosan exhibits antibacterial and hemostatic properties, which can reduce infection and facilitate blood clotting. In contrast, pectin is characterized by its soothing nature and non-adherence to new tissue, resulting in less painful dressing changes due to its low cell adhesion, preventing damage to regenerating layers. A recent study developed a bioadhesive pectin-chitosan hydrogel containing ciprofloxacin (CIP) for the treatment of infected wounds [28]. The hydrogel exhibited a high-water content and retention capacity, establishing an optimal moist environment, and demonstrated no cytotoxicity or hemolysis *in vitro* [28]. The incorporation of a dopamine-modified micellar CIP carrier enhanced drug retention within the gel and extended its release at the wound site [28]. *In vitro* and *in vivo* studies indicated that the CIP-loaded Ch-Pec hydrogel effectively eradicated bacteria and enhanced wound healing, highlighting its potential as an anti-infective wound dressing [28].

Film dressings and membranes: Thin flexible films composed of pectin and chitosan can function as wound coverings. These can be utilized similarly to bandages and are especially effective for superficial or partial-thickness wounds. The mechanical robustness of Ch-Pec films typically exceeds that of pectin-only films, attributable to the reinforcing effect of chitosan. Muthulakshmi et al. [35] created a biodegradable wound dressing using a curcumin-incorporated pectin-chitosan film. The incorporation of chitosan enhanced the tensile strength and thermal stability of the pectin film while decreasing its vapor transmission, thereby promoting moisture retention for wound care. Curcumin, recognized for its natural anti-inflammatory and antimicrobial properties, further augmented these effects. The optimized film, referred to as PC4/CC1.5, demonstrated significant antibacterial activity against both Gram-positive and Gram-negative bacteria. This example demonstrates the dual role of a pectin-chitosan film as both a delivery vehicle for the therapeutic agent curcumin and a functional dressing that combats infection and promotes tissue repair [36].

Other studies have similarly incorporated bioactives such as honey, propolis, silver nanoparticles, or growth factors into Ch-Pec films to improve wound healing. Propolis-loaded quaternate chitosan/pectin films demonstrated extensive antibacterial properties and enhanced *in vitro* healing assessments [46]. Demchenko et al. [37] investigated the impact of interpolyelectrolyte complexes (IPEC) composed of chitosan, pectin, and starch on the structural and antimicrobial characteristics of silver-containing nanocomposites. Their findings indicate that the pectin-Ag-chitosan complex exhibits the highest antimicrobial activity [38].

Porous Sponges: As noted, freeze-dried Ch-Pec sponges can serve as absorbent wound dressings, particularly for exuding wounds or as fillers for deep cavity wounds. A porous sponge can absorb fluid, conform to the wound bed, and later be resorbed or easily removed. Pectin's presence can modulate the chitosan sponge's pH responsiveness and reduce its shrinkage when hydrated. A composite sponge with 30% pectin (rest chitosan) was reported to have high porosity (~90%), good fluid absorption, and still retain the intrinsic antimicrobial activity of chitosan [31]. *In vivo*, such sponges supported faster wound closure and better tissue organization compared to commercial dressings, as pectin improved the biocompatibility of the scaffold. Furthermore, since pectin can be ionically crosslinked with calcium, some wound sponges include calcium to add structural integrity and also help in blood clotting (Ca^{2+} is involved in coagulation).

Antibacterial and anti-biofilm action: Wound infections (biofilms) are a major impediment to healing. Chitosan is inherently antimicrobial due to its polycationic nature, disrupting microbial cell membranes. The combination with pectin often does not diminish this; in fact, pectin-chitosan materials maintain strong antibacterial effects [46]. The PCPC implant coating study, while aimed at implants, is relevant: coatings with a higher chitosan ratio showed significant zones of inhibition against *Staphylococcus aureus* and *Escherichia coli*, indicating the composite can prevent bacterial colonization on surfaces [4]. In wound dressings, this translates to a material that can reduce bioburden in the wound. The CIP-loaded hydrogel mentioned earlier provided sustained local antibiotic release to kill bacteria over time [46]. Thus, a pectin-chitosan dressing can tackle infection both *intrinsically* (chitosan's action) and via *payload* (released antiseptics or antibiotics). This dual action is particularly valuable against resilient biofilms in chronic wounds.

5.2 Tissue engineering and regenerative medicine

In tissue engineering, scaffolds made of pectin-chitosan have been explored to support the growth of new tissue (bone, cartilage, skin, etc.) or as 3D culture systems for tissue models. The rationale is that combining chitosan and pectin yields a matrix more akin to natural extracellular matrix (ECM) in its complexity and properties than either polymer alone [1]. Important parameters for tissue engineering include porosity, cell-compatibility, and suitable mechanical strength—all of which Ch-Pec scaffolds allow to be modified.

3-D scaffolds for cell growth: Freeze-dried Ch-Pec scaffolds, as described, offer porous structures on which cells may connect and grow. With MTT tests proving excellent viability, Coimbra et al. [38] showed that a pectin-chitosan polyelectrolyte complex scaffold facilitated adhesion and differentiation of human osteoblasts. Pectin, a plant polysaccharide not typically found in mammalian ECM, did not affect the cells; rather, the hybrid scaffold encouraged osteogenic activity, perhaps because chitosan's amino groups give a favorable surface and pectin contributed to a softer, more hydrated matrix [39]. Birch et al. [47] also evaluated human mesenchymal stem cells (hMSCs) acidic-synthesized Ch-Pec hydrogels. Although initial cell adhesion was low—probably helpful for non-adherent uses like wound dressings—the general cell proliferation increased with minor changes (rinsing the gels to eliminate extra acid). This indicates that by optimizing fabrication (e.g., residual acid removal, adding cell-adhesive peptides), Ch-Pec hydrogels can be quite hospitable to cells. In fact, modifying pectin or chitosan with RGD peptides (arginine-glycine-aspartic acid sequences that promote cell adhesion) dramatically enhances cell attachment.

Thermosensitive and injectable scaffolds: For tissue engineering *in vivo*, an injectable scaffold that gels *in situ* is highly attractive (it can fill irregular defects and be delivered minimally invasively). Thermosensitive Ch-Pec hydrogels have been studied for this purpose. *Morello et al.* [48] reported a thermo-responsive chitosan/pectin hydrogel that could sustain long-term culture of tumor spheroids [48] – this suggests a mild gelation process compatible with live cells. Another group formulated an injectable Ch-Pec hydrogel with cellulose nanocrystal reinforcement to achieve an optimal gel for cartilage tissue engineering, which was shear-thinning and injectable, then rapidly re-solidified in the tissue defect [42]. These materials exemplify how Ch-Pec systems can be combined with other components (nanocellulose, bioactive glass, etc.) to meet specific mechanical or biological requirements of a tissue scaffold.

Bone and Cartilage Regeneration: Chitosan has been widely used in bone tissue engineering, and adding pectin can introduce osteoconductive cues and enhance printability or gelation. Recent work [1] cites that RGD-functionalized pectin gels, or pectin crosslinked with calcium carbonate, have been successfully used as bioinks for 3D bioprinting bone tissue models. The ability of pectin to gel with $CaCO_3$ was exploited to print scaffolds that subsequently supported neural cell cultures in a 3D brain model showing pectin's utility even beyond typical musculoskeletal applications. For cartilage, chitosan provides a cartilage-like glycosaminoglycan analog, and pectin can improve the viscoelastic properties [1].

Pectin-chitosan scaffolds in tissue engineering allow one to leverage chitosan's structural and antimicrobial features with pectin's gelling and biofunctional attributes. They can be processed into many forms (bulk hydrogels, foams, injectable gels, and 3D-printed structures) and can be enriched with cell-signaling motifs. Early studies show they support cell viability and even differentiation, although surface modifications (e.g., RGD) can greatly improve cell attachment due to pectin and chitosan's lack of native mammalian cell-binding sequences. The composite nature of Ch-Pec scaffolds often results in a more ECM-mimetic environment than single-polymer scaffolds, which can translate to better tissue regeneration outcomes.

5.3 Encapsulation of bioactives and biotechnological uses

Beyond pharmaceuticals, Ch-Pec combinations are utilized to encapsulate and transport a range of bioactive chemicals – including nutraceuticals, enzymes, probiotics, flavors, and cells – in pharmaceutical, food, and biotech settings. For delicate biological loads, their delicate formulation conditions (typically room temperature and watery) and generally recognized as safe (GRAS) status make them perfect:

Encapsulation of probiotics and nutritional agents. Extensive research has gone on pectin-chitosan hydrogel beads and capsules for the digestive system protection of delicate components. To increase the survival rate in stomach acid, probiotic bacteria, such as *Lactobacillus* spp., have been encapsulated in calcium pectate beads covered with chitosan [19]. The chitosan coating prevents breakdown until the intestines, therefore providing a high viable count of probiotics to the colon. Acevedo et al. emphasized that pectin's mild gelation and environment fit cells, thus it is perfect for probiotic encapsulation [19]. They solved a major problem in oral probiotic formulations by adding a chitosan layer, therefore preventing the low pH-induced viability loss of probiotics in the stomach. Ch-Pec matrices have similarly captured vitamins, antioxidants, and essential oils. Investigating essential oil-loaded pectin/chitosan nanoparticles revealed that the system could preserve the volatile oil and then release it in a regulated manner, therefore exhibiting anticancer properties on MCF-7 cells [43]. Ch-Pec films and coatings are being investigated in the food industry to package and distribute functional ingredients, for example, Tong et al. [49] created a biodegradable active food packaging film with cinnamic acid by combining sodium alginate, pectin, and chitosan, which shows slowed oxidation and microbial growth on food [50]. Like pH-triggered taste release in oral care products, pectin-chitosan microcapsules have also been utilized for disguising the tastes or odors of substances and for targeted release.

Enzyme and cell immobilization: In biotechnology, pectin-chitosan beads are immobilization mats for enzymes or yeast cells in bioreactors. Mild aeration and substantial water content facilitate the encapsulation of biomolecules while preventing their denaturation. For ethanol fermentation, for example, pectin-chitosan co-gels have been employed to immobilize yeast therefore enabling reuse of the biocatalyst and simple separation from the result. Such beads' porous shape keeps the enzyme or cells while allowing substrate and product diffusion. Furthermore, the polyelectrolyte character helps stabilize certain enzymes; for instance, a positively charged enzyme can attach itself in the gel by interacting with COO^- groups of pectin. Strong electrostatic binding and microenvironment given by pectin (which can change local pH) has shown research that pectin-chitosan matrices increased the operational stability of immobilized enzymes like β-galactosidase, compared to other polymers [1, 51].

Biologically active polymers: Though chitosan is recognized for its antibacterial and hemostatic properties, pectin itself has some documented biological activity (immunomodulatory, cholesterol-lowering, etc.). Sometimes the composite itself is the bioactive agent in some uses. Pectin, for instance, especially modified pectin, can have anticancer action by causing death or blocking galectin in tumor cells. Chitosan can induce reactions of wound healing. Therefore, a Ch-Pec material might offer a therapeutic advantage beyond only serving as a passive carrier. Reviewing pectin-containing biomaterials, Eivazzadeh-Keihan et al. [52] found anticancer, antioxidant, and anti-inflammatory effects. These benefits might be combined in a composite; for example, a pectin-chitosan wound dressing not only provides extra medications but also naturally lowers inflammation and scavenges free radicals [53].

Agriculture and environmental uses: Although our focus is biomedical, pectin-chitosan formulations are under investigation in agricultural biotechnology (e.g., as carriers for pesticides or as seed coatings that degrade into nutrients) and in environmental engineering (e.g., pectin-chitosan gels for heavy metal ion adsorption from water, given both have functional groups that bind metal). These underline the general use of Ch-Pec complexes in several fields.

Ch-Pec systems are essentially strong encapsulation platforms that release sensitive bioactives under specified stimuli (pH, enzymes, time) while preserving them. They are gentle enough for sensitive biomolecules and edible and safe for food or oral pills as they mix the advantages of two natural polymers. **Table 1** shows increased stability and targeted distribution by listing a few instances of bioactive encapsulation (probiotics, nutraceuticals) and their results.

5.4 Antimicrobial and biofilm materials

Chitosan and pectin both help to improve the antibacterial characteristics of composite materials. Chitosan's mechanisms include disrupting microbial membranes and chelating metals required for microbial growth [46], whereas pectin, particularly when used as oligogalacturonides, can inhibit bacterial adhesion and biofilm formation (and can be metabolized by gut flora, producing pathogen-suppressing acids) [54]. Pectin-chitosan composites have been used to create antimicrobial films, coatings, and wound dressings.

Antibacterial films and coatings: We described how pectin-chitosan wound films and implant coatings exhibit antibacterial properties. The Science Reports 2023 research on PCPC (pectin-chitosan) coatings for titanium implants is a good example. They tested various ratios; coatings with higher chitosan content (groups 3, 4, 5 corresponding to 1:2, 1:3, etc.) exhibited clear zones of inhibition against bacteria, whereas uncoated titanium had none. Importantly, all tested PCPC ratios were non-cytotoxic to human cells (osteoblast-like cells), indicating they can ward off bacteria without harming host tissue [4]. This opens the door to using such coatings on medical devices to reduce infections (e.g., dental implants and catheters). For food packaging, antimicrobial Ch-Pec films have been used to extend shelf life by preventing microbial growth on food surface [40]. The incorporation of natural antimicrobials (e.g., essential oils, nisin, and silver ions) into pectin-chitosan films further boosts their efficacy. A study demonstrated that nisin, an antimicrobial peptide suitable for food applications, encapsulated within a Ch-Pec nanoparticle exhibited improved stability and an extended antibacterial effect [26]. Embedding nanoparticles in a film or coating facilitates the sustained release of antimicrobial agents, thereby preserving surface hygiene over an extended period.

Hydrogels that inhibit biofilm formation: Chronic wounds and implants often exhibit biofilm formation, which poses greater challenges for elimination compared to planktonic bacteria. Pectin-chitosan hydrogels can be formulated with anti-biofilm agents, such as dispersin B enzyme or nitric oxide donors, to effectively target biofilms locally. The CIP-loaded hydrogel, for instance, maintained an effective antibacterial concentration at the wound for an extended period, which is crucial for penetrating and disrupting biofilms [46]. Moreover, chitosan is known to inhibit *S. aureus* and *Pseudomonas aeruginosa* biofilm formation by preventing initial adhesion and by detaching biofilm matrix through charge interactions [41].

Antifungal properties: Chitosan is also antifungal (effective against *Candida*, *Aspergillus*, etc.) [43]. Pectin-chitosan films have shown activity against fungal pathogens, important for applications like oral thrush treatments or antifungal packaging. For example, a study reported that adding chitosan to pectin films inhibited *Candida albicans* growth, whereas pectin-alone films did not (owing to chitosan's cationic nature disrupting fungal cells). This broad-spectrum antimicrobial effect (covering Gram+ bacteria, Gram– bacteria, and fungi) makes Ch-Pec materials broadly protective.

Pectin-chitosan systems are great examples of naturally occurring antimicrobial compounds that may either act as carriers of other antimicrobial agents or naturally resist microorganisms. By offering local antimicrobial activity, they lessen the requirement for systemic antibiotics and meet the rising demand for biocompatible infection control methods.

Pectin-chitosan compounds are important because of their many uses, which include antibacterial coatings, tissue engineering, and medication delivery. Whether it is pH-triggered release, moisture retention, mechanical support, or bioadhesion, each application makes use of a distinct feature of the Ch-Pec composite.

6. Other biopolymer systems

Finally, it is critical to situate pectin-chitosan composites within other natural polymer combinations (alginate-chitosan, chitosan-hyaluronan, chitosan-gelatin, etc.) and address their special benefits and drawbacks:

6.1 Advantages of pectin-chitosan systems

Broad pH Stability Range: Pure chitosan is only soluble (and usable) in acidic environments (pH < 6). When complexed with pectin, the composite can remain stable across a broader pH range – pectin effectively "stabilizes" chitosan by binding it and preventing its premature dissolution. For instance, a study noted that while chitosan dissolves in acid, adding pectin keeps the complex intact at lower pH values [30]. This is advantageous over systems like chitosan-gelatin which tend to dissolve in physiological conditions unless chemically crosslinked. It means Ch-Pec formulations can survive stomach acid (unlike alginate-chitosan which may disintegrate in very low pH as alginate becomes alginic acid).

Colon degradability (targeted delivery): Unlike some other polyanions (e.g., alginate or hyaluronic acid), pectin is readily degraded by human colonic microbiota (pectinases, glycosidases) [19]. This makes Ch-Pec systems particularly suited for colon-targeting – the pectin network will break down in the colon, ensuring payload release or scaffold degradation *in situ*. Alginate-chitosan systems have been used similarly, but alginate is not as specifically degraded by colonic flora (it mainly ion-exchanges in different pH). Pectin-chitosan stands out for applications requiring microbially-triggered disintegration, such as colon-specific drug delivery or gut-active nutraceutical release.

Mucoadhesion and bioadhesion: Many natural polyelectrolyte complexes (e.g., chitosan-alginate) are mucoadhesive due to charge interactions. Chitosan-pectin is strongly mucoadhesive because chitosan's cationic charges bind to the negatively charged mucins on mucosal surfaces. Pectin also has mucoadhesion properties related to hydrogen bonding between its free acid groups and mucin. Additionally, pectin, being a viscous polysaccharide, can increase the residence time [44]. Compared to synthetic polymers or even chitosan-carboxymethylcellulose, Ch-Pec shows impressive adhesion in GI and buccal environments (as evidenced by buccal film studies giving high bioavailability. This means Ch-Pec systems often outperform purely synthetic delivery systems (e.g., PLA/PLGA nanoparticles) in keeping drugs at the absorption site longer, which can improve therapeutic efficacy.

No need for chemical crosslinkers: Formulations like chitosan-collagen or chitosan-hyaluronan often require crosslinking agents (e.g., glutaraldehyde and genipin) to form a stable hydrogel or film. Pectin-chitosan, however, self-crosslinks ionically, avoiding potentially toxic reagents [1, 20]. This simplifies preparation and improves biocompatibility. For example, a chitosan-gelatin scaffold might need genipin to prevent it from dissolving in culture, whereas a chitosan-pectin scaffold can hold without additional chemicals due to the polyelectrolyte network. The "green" preparation process (just mixing in water, sometimes with salt or mild heat) is a clear advantage for biomedical use.

Customizability via pectin type: As discussed, by choosing HMP vs. LMP vs. amidated pectin, one can tune the composite's properties. This flexibility is somewhat unique – other polyanions like alginate mostly vary by M/G ratio and molecular weight but not in fundamentally different categories like HMP vs. LMP. Amidated pectin-chitosan films for instance showed better mechanical properties than those with non-amidated pectin [20]. Such fine control is beneficial for optimizing a product for a specific need (e.g., a stronger film for dressing vs. a faster-degrading gel for drug release).

Natural and generally recognized as safe (GRAS): Both pectin and chitosan are largely regarded as safe, non-immunogenic, and biocompatible. Competing systems like chitosan-synthetic polymer (e.g., chitosan-PVA blends) have the downside of the synthetic component possibly not degrading or having toxic monomers. Chitosan-pectin fully degrades into harmless sugars and amino-sugars. This is a regulatory and safety advantage, and also environmentally friendly [45].

The table below compares pectin-chitosan (Ch-Pec) systems with other common chitosan-based biopolymer composites – chitosan-alginate (Ch-Alg), chitosan-gelatin (Ch-Gel), and chitosan-synthetic polymer combinations (e.g., Ch-PLGA/PCL/PVA). Key properties are compared in terms of advantages and challenges for each system (**Table 2**).

6.2 Challenges and considerations

Reproducibility and material variability: A classic challenge with pectin-chitosan is the variability in pectin's structure from batch to batch (different plant sources or extraction methods). While standardized pectin exists now, they still have more lot-to-lot variation than pure synthetic polymers [19]. This can affect gel strength or drug release profiles. Careful characterization (DM, molecular weight) of each pectin batch is needed to ensure consistent performance – something less of an issue in, say, alginate (which also has variability, but a more established supply for medical grade).

pH sensitivity (limitations in neutral pH): The same pH-responsive property that is an advantage for targeted release can be a disadvantage for other uses. Specifically, in physiological pH (~7.4) or higher, the ionic crosslinks in Ch-Pec weaken as chitosan loses charge [1]. This means an uncross-linked Ch-Pec hydrogel might not be stable in blood or tissue fluid unless reinforced (either covalently or by secondary interactions). Competing systems like chitosan-researchers address this by slight covalent crosslinking or by formulating the hydrogel to remain in a mildly acidic microenvironment. But it remains a challenge if one wanted to use a Ch-Pec vehicle in the bloodstream or in long-term implants in alkaline conditions.

Property	Chitosan-Pectin (Ch-Pec)	Chitosan-Alginate (Ch-Alg)	Chitosan-Gelatin (Ch-Gel)	Chitosan-Synthetic (PLGA/PCL/PVA)
pH Stability	pH responsive. Forms polyelectrolyte complexes: stable through GI transit, mainly releasing in colon [55, 56]	pH sensitive. Protects in stomach, dissolves in small intestine [57]	Not inherently stable. Requires crosslinking [58]	High pH stability. Synthetic polymers resist pH changes, degrade by hydrolysis [56]
Biodegradability	Fully biodegradable by natural enzymes/microflora [56, 58]	Biodegradable, but dependent on microbial presence [56]	Enzymatically biodegradable [58]	Polymer-dependent biodegradation; PLGA/PCL degrade slowly [56]
Mechanical Strength	Moderate to good. Pectin enhances elasticity and strength [59].	Moderate. Alginate's swelling reduces mechanical stability [59].	Adjustable through chemical or physical crosslinking [58].	High. Synthetic polymers provide excellent durability [50, 58].
Mucoadhesiveness	High. Strong adhesion *via* ionic and hydrogen bonds [56, 57].	High. Good mucoadhesion through alginate-chitosan interaction [56].	Moderate. Rely mainly on chitosan; gelatin less active.	Low unless chitosan coating is applied [56].
Crosslinking Needs	Low. Simple ionic complexation [51].	Low. Ionic gelation sufficient [51].	High. Require chemical crosslinkers [53, 58].	Moderate. Depending on polymer, some need no crosslinkers [58].
Release Control	Colon-triggered & sustained *via* enzymatic breakdown [58].	pH-dependent release in upper intestine [57].	Diffusion/erosion controlled sustained release [58].	Sustained and programmable release based on polymer degradation [54, 56].
Colon-Targeted Delivery	High. Effective colon-specific delivery [36, 55, 58] .	Moderate/Low. Tends to release earlier unless modified [53].	Low. No intrinsic colon targeting.	Low. Needs additional coating for colon-specific delivery [53].
Bioactivity of Matrix	Bioactive: antimicrobial, prebiotic, anti-inflammatory [56].	Mildly bioactive: mainly chitosan's contribution [56].	Highly bioactive: supports tissue growth and healing [58].	Minimal. Synthetic polymers are inert; chitosan improves bioactivity slightly [56].
Biocompatibility	Excellent. Natural, safe degradation products [56, 59].	Excellent. Extensively used in medical application [56].	Excellent. Gelatin and chitosan are highly biocompatible [58].	Generally high. Synthetic polymers are well tolerated [56].
Reproducibility	Moderate. Natural polymer variability [56].	Moderate. Needs stringent quality control [56].	Moderate. Variability from gelatin/chitosan source.	High (materials)/Moderate (process) [56].

Property	Chitosan-Pectin (Ch-Pec)	Chitosan-Alginate (Ch-Alg)	Chitosan-Gelatin (Ch-Gel)	Chitosan-Synthetic (PLGA/PCL/PVA)
Scalability	High. Materials are abundant; methods are scalable [56]	High. Industrially scalable processes exist [56]	Moderate. Crosslinking and processing steps are complex [58]	Moderate. Scalable but capital intensive [57]

Table 2.
Comparative evaluation of chitosan-pectin vs. other chitosan biopolymer composites.

Mechanical strength: Pure polyelectrolyte complexes of chitosan-pectin 2019 review by Martău et all tends to be softer and more brittle when dry compared to some synthetic polymer blends [34]. For instance, chitosan-PEG or chitosan-PCL composites can achieve higher tensile strength (albeit losing biodegradability). Chitosan-pectin films, while improved over pectin alone, may still require reinforcement (e.g., a backing layer or nanoparticles) for certain high-stress applications. The wet strength of Ch-Pec hydrogels is often lower than, say, chitosan-silk fibroin hybrid hydrogels or chitosan-PLA interpenetrating networks which benefit from strong covalent networks. However, semi-IPN approaches (e.g., a lightly crosslinked chitosan network interpenetrated by pectin) can boost mechanical properties. Researchers have also added fillers like clay nanosheets or cellulose nanofibers to Ch-Pec to reach the desired stiffness [32].

Fast swelling and potential burst release: Polyelectrolyte hydrogels can imbibe water rapidly and sometimes release a large fraction of the loaded drug in an initial burst. Without careful formulation, a pectin-chitosan particle might dump a drug upon encountering physiological salt (which screens charges and causes the gel to loosen). Other systems like chitosan-tripolyphosphate NPs have similar issues, but can be more easily tuned by crosslinker amount. In Ch-Pec, achieving a very *tight* network may require high charge density (high DA chitosan and low DM pectin) and optimal stoichiometry. If not, you might see an initial burst release or gradual erosion that is too quick for long-term delivery. This is a challenge shared with alginate-chitosan beads as well.

Interactions with charged biomolecules: While mucoadhesion is a boon, it also means Ch-Pec materials can nonspecifically bind other molecules. For instance, if a Ch-Pec nanoparticle is in the bloodstream, it might adsorb proteins (opsonization) readily due to its charged surface, potentially accelerating clearance by the immune system. This is a general issue with cationic nanocarriers (chitosan, PEI, etc.). Some strategies like PEGylation or coating with neutral polymers have been tested (e.g., PEGylated Ch-Pec NPs in a colon delivery study [60]. Indeed, one study created PEGylated pectin-chitosan NPs that showed improved stability and cellular uptake for colon cancer therapy [60]. Thus, while Ch-Pec on its own is bioadhesive, in situations where that is detrimental, an extra surface modification may be needed.

Comparison to Alginate-Chitosan: Alginate-chitosan PECs are very widely studied (alginate being another polyanionic polysaccharide) [61]. Both systems are fairly similar in concept, but pectin has an edge in biodegradability (alginate is not enzymatically degraded in mammals and can persist longer). On the other hand, alginate gels often have more uniform well-defined crosslinking zones (G-blocks with Ca^{2+}) and there are FDA-approved alginate wound dressings and embolic beads, indicating strong precedents. Pectin-chitosan is still catching up in terms of commercialization. However, pectin-chitosan might prove superior for oral applications where complete breakdown and fermentation is desired (alginate can cause some GI side effects if not fully eliminated). A 2019 review by Martău et al. compared chitosan, alginate, and pectin in biomedical use and highlighted that all three are biocompatible and biodegradable, but chitosan-pectin combinations showed the best balance of bioadhesiveness and degradability [34].

Comparison to chitosan-synthetic polymer systems: Synthetic polymers (PLGA, PCL, PVA, etc.) offer strength and controlled degradation but often lack biological interactions. Chitosan-pectin's big advantage is that it is *bioactive* (cells can respond to oligosaccharides from pectin, and chitosan's fragments can stimulate healing). However, synthetic blends often excel in mechanical strength and precision of release (due to well-defined degradation kinetics). The choice may come down to application: for a temporary implant that needs high load bearing (like a bone screw), a chitosan-PCL might be chosen over Ch-Pec; but for a bioresorbable tissue scaffold or drug depot that should integrate with tissue and then vanish, Ch-Pec is very attractive.

Ultimately, pectin-chitosan composites have a distinctive niche: they are totally natural, naturally pH-responsive, and allow colon-specific or mucoadhesive distribution that many other methods cannot do as simple. Mostly linked to stability under physiological circumstances and consistency, they provide difficulties; yet, continuous research (e.g., chemical changes and hybrid networks) is solving these problems.

7. Conclusion

In the past decade, pectin-chitosan composite biomaterials have advanced significantly, proving to be a versatile platform for drug delivery, regenerative medicine, and biotechnology. By combining pectin (a plant-derived polyanion with excellent gel-forming and biodegradation properties) and chitosan (a marine-derived polycation known for its bioadhesiveness and antimicrobial activity), researchers have created ionically crosslinked networks that do not require toxic crosslinkers and respond smartly to environmental triggers (pH, enzymes, ionic strength). This chapter describes those various types of pectin (high-methoxy, low-methoxy, amidated) that can be paired with chitosan to tune the composite's behavior – from robust sustained-release nanoparticles to fast-gelling injectable hydrogels. Key interaction mechanisms (electrostatic complexation) underpin the self-assembling nature of these materials, while formulation innovations (layer-by-layer films, core-shell particles, freeze-dried scaffolds) have broadened their applicability.

Applications of pectin-chitosan systems are wide-ranging. In drug delivery, Ch-Pec systems achieve site-specific release (notably colon targeting for inflammatory bowel disease or colon cancer therapy) and controlled release profiles for oral, buccal, ocular, and transdermal routes, often improving therapeutic outcomes and reducing side effects.

Preclinical studies in rats, rabbits, and *in vitro* human cell cultures have reinforced the biocompatibility and efficacy of these systems – healing wounds faster, delivering drugs more effectively, and not eliciting toxic responses. While clinical translation is still forthcoming, the outlook is positive given that both pectin and chitosan are familiar, safe materials already used in humans. With continued research focusing on addressing challenges (like ensuring stability *in vivo*, scaling manufacturing, and standardizing polymer quality), it is likely that pectin-chitosan composite therapies will move into clinical trials. They offer a natural, sustainable, and multifunctional alternative to synthetic biomaterials, aligning well with the current drive toward greener and patient-tailored biomedical technologies.

In conclusion, the past decade has firmly established pectin-chitosan combinations as a preferred biopolymer system for innovators focused on developing advanced delivery carriers, regenerative scaffolds, and bioactive coatings. The synergy of pectin and chitosan – one providing a negative charge and enzymatic degradability, the other providing a positive charge and mucoadhesion – is the cornerstone of its success, yielding composites that are more than the sum of their parts. Future research may explore new pectin modifications, chitosan derivatives, or hybridizing Ch-Pec with inorganic nanoparticles to further expand functionality (such as magnetic or stimulus-responsive Ch-Pec gels). The fundamental research reviewed in this chapter sets the stage for such innovations. Given their promising performance and inherent safety, pectin-chitosan biomaterials are poised to make the leap from bench to bedside, contributing to improved therapies in drug delivery, wound care, and beyond.

Acknowledgements

This work was financially supported by "Carol Davila" University of Medicine and Pharmacy Bucharest, Romania, through Contract no. CNFIS-FDI-2025-F-0646.

Conflict of interest

We, the authors of this chapter: Lăcrămioara Popa, Mihaela Violeta Ghica, Cristina-Elena Dinu-Pîrvu, and Irina-Alexandra Dumitrescu, declare no conflicts of interest.

Author details

Lăcrămioara Popa[1,2], Mihaela Violeta Ghica[1,2]*, Cristina-Elena Dinu-Pîrvu[1,2] and Irina-Alexandra Dumitrescu[1,2]

1 Faculty of Pharmacy, Carol Davila University of Medicine and Pharmacy Bucharest, Romania

2 Innovative Therapeutic Structures Research and Development Center (InnoTher), "Carol Davila" University of Medicine and Pharmacy, Bucharest, Romania

*Address all correspondence to: mihaela.ghica@umfcd.ro

References

[1] Morello G, De Iaco G, Gigli G, Polini A, Gervaso F. Chitosan and pectin hydrogels for tissue engineering and in vitro modeling. Gels. 2023;**9**(2):132. DOI: 10.3390/gels9020132

[2] Kocaağa B et al. Developing multifunctional pectin-based hydrogel for wound dressing: In silico, in vitro and in vivo evaluation. European Polymer Journal. 2024;**216**:113280. DOI: 10.1016/j.eurpolymj.2024.113280

[3] Mittal H et al. Recent progress in the structural modification of chitosan for applications in diversified biomedical fields. European Polymer Journal. 2018;**109**:402-434. DOI: 10.1016/j.eurpolymj.2018.10.013

[4] Alsharbaty MHM, Naji GA, Ali SS. Exploring the potential of a newly developed pectin-chitosan polyelectrolyte composite on the surface of commercially pure titanium for dental implants. Scientific Reports. 2023;**13**(1):22203. DOI: 10.1038/s41598-023-48863-2

[5] Tian L, Singh A, Singh AV. Synthesis and characterization of pectin-chitosan conjugate for biomedical application. International Journal of Biological Macromolecules. 2020;**153**:533-538. DOI: 10.1016/j.ijbiomac.2020.02.313

[6] Wasupalli GK, Verma D. Thermosensitive injectable hydrogel based on chitosan-polygalacturonic acid polyelectrolyte complexes for bone tissue engineering. Carbohydrate Polymers. 2022;**294**:119769. DOI: 10.1016/j.carbpol.2022.119769

[7] Amirian J et al. In-situ crosslinked hydrogel based on amidated pectin/oxidized chitosan as potential wound dressing for skin repairing. Carbohydrate Polymers. 2021;**251**:117005. DOI: 10.1016/j.carbpol.2020.117005

[8] Mallick SP, Singh BN, Rastogi A, Srivastava P. Design and evaluation of chitosan/poly(l-lactide)/pectin based composite scaffolds for cartilage tissue regeneration. International Journal of Biological Macromolecules. 2018;**112**:909-920. DOI: 10.1016/j.ijbiomac.2018.02.049

[9] Soubhagya AS, Moorthi A, Prabaharan M. Preparation and characterization of chitosan/pectin/ZnO porous films for wound healing. International Journal of Biological Macromolecules. 2020;**157**:135-145. DOI: 10.1016/j.ijbiomac.2020.04.156

[10] Andishmand H, Tabibiazar M, Mohammadifar MA, Hamishehkar H. Pectin-zinc-chitosan-polyethylene glycol colloidal nano-suspension as a food grade carrier for colon targeted delivery of resveratrol. International Journal of Biological Macromolecules. 2017;**97**:16-22. DOI: 10.1016/j.ijbiomac.2016.12.087

[11] Long J, Etxeberria AE, Nand AV, Bunt CR, Ray S, Seyfoddin A. A 3D printed chitosan-pectin hydrogel wound dressing for lidocaine hydrochloride delivery. Materials Science and Engineering: C. 2019;**104**:109873. DOI: 10.1016/j.msec.2019.109873

[12] Wang Y et al. Fabricating pectin and chitosan double layer coated liposomes to improve physicochemical stability of beta-carotene and alter its gastrointestinal fate. International Journal of Biological Macromolecules. 2023;**247**:125780. DOI: 10.1016/j.ijbiomac.2023.125780

[13] Alsharbaty MHM, Naji GA, Ghani BA, Schagerl M, Khalil MA, Ali SS. Cytotoxicity and antibacterial susceptibility assessment of a newly developed pectin–chitosan polyelectrolyte composite for dental implants. Scientific Reports. 2024;**14**(1):16968. DOI: 10.1038/s41598-024-68020-7

[14] Niu H, Hou K, Chen H, Fu X. A review of sugar beet pectin-stabilized emulsion: Extraction, structure, interfacial self-assembly and emulsion stability. Critical Reviews in Food Science and Nutrition. 2024;**64**(3):852-872. DOI: 10.1080/10408398.2022.2109586

[15] EFSA Panel on Food Additives and Nutrient Sources added to Food (ANS) et al. Re-evaluation of pectin (E 440i) and amidated pectin (E 440ii) as food additives. EFSA Journal. 2017;**15**(7):e04866. DOI: 10.2903/j.efsa.2017.4866

[16] Khotimchenko M. Pectin polymers for colon-targeted antitumor drug delivery. International Journal of Biological Macromolecules. 2020;**158**:1110-1124. DOI: 10.1016/j.ijbiomac.2020.05.002

[17] Fernández-Hervás MJ, Fell JT. Pectin/chitosan mixtures as coatings for colon-specific drug delivery: An in vitro evaluation. International Journal of Pharmaceutics. 1998;**169**(1):115-119. DOI: 10.1016/S0378-5173(98)00114-8

[18] Yang X, Nisar T, Liang D, Hou Y, Sun L, Guo Y. Low methoxyl pectin gelation under alkaline conditions and its rheological properties: Using NaOH as a pH regulator. Food Hydrocolloids. 2018;**79**:560-571. DOI: 10.1016/j.foodhyd.2017.12.006

[19] Morales E et al. Pectin–chitosan hydrogel beads for delivery of functional food ingredients. Food. 2024;**13**(18):2885. DOI: 10.3390/foods13182885

[20] Potaś J, Wilczewska AZ, Misiak P, Basa A, Winnicka K. Optimization of multilayer films composed of chitosan and low-methoxy amidated pectin as multifunctional biomaterials for drug delivery. International Journal of Molecular Sciences. 2022;**23**(15):8092. DOI: 10.3390/ijms23158092

[21] Jonassen H, Treves A, Kjøniksen A-L, Smistad G, Hiorth M. Preparation of ionically cross-linked pectin nanoparticles in the presence of chlorides of divalent and monovalent cations. Biomacromolecules. 2013;**14**(10):3523-3531. DOI: 10.1021/bm4008474

[22] Assifaoui A, Hayrapetyan G, Gallery C, Agoda-Tandjawa G. Exploring techno-functional properties, synergies, and challenges of pectins: A review. Carbohydrate Polymer Technologies and Applications. 2024;**7**:100496. DOI: 10.1016/j.carpta.2024.100496

[23] Aranaz I et al. Chitosan: An overview of its properties and applications. Polymers (Basel). 2021;**13**(19):3256. DOI: 10.3390/polym13193256

[24] Shitrit Y, Davidovich-Pinhas M, Bianco-Peled H. Shear thinning pectin hydrogels physically cross-linked with chitosan nanogels. Carbohydrate Polymers. 2019;**225**:115249. DOI: 10.1016/j.carbpol.2019.115249

[25] Giacomazza D, Sabatino MA, Catena A, Leone M, San Biagio PL, Dispenza C. Maltose-conjugated chitosans induce macroscopic gelation of pectin solutions at neutral pH. Carbohydrate Polymers. 2014;**114**:141-148. DOI: 10.1016/j.carbpol.2014.08.014

[26] Silant'ev VE et al. Rational design of pectin–chitosan polyelectrolyte

nanoparticles for enhanced temozolomide delivery in brain tumor therapy. Biomedicine. 2024;**12**(7):1393. DOI: 10.3390/biomedicines12071393

[27] Yin B et al. Engineering multifunctional dynamic hydrogel for biomedical and tissue regenerative applications. Chemical Engineering Journal. 2024;**487**:150403. DOI: 10.1016/J.CEJ.2024.150403

[28] Cheng K-C et al. Theophylline-loaded pectin/chitosan hydrochloride submicron particles prepared by spray drying with a continuous feeding ultrasonic atomizer. Polymers (Basel). 2022;**14**(21):4538. DOI: 10.3390/polym14214538

[29] Pachaleva J, Gruskiene R, Bockuviene A, Sereikaite J. The application of pectin and chitooligosaccharides for the preparation of nisin-loaded nanoparticles with different surface charges. Cellulose. 2023;**30**(14):8985-8996. DOI: 10.1007/s10570-023-05380-9

[30] Boughanmi R, Oelmann M, Steinbach C, Schwarz S. Sustainable polyelectrolyte complexes of pectin and chitosan as adsorbents for heavy metal ions from surface water. Journal of Polymer Science. 2025;**63**(1):133-145. DOI: 10.1002/pol.20240442

[31] Kraskouski A et al. Design of highly porous materials based on chitosan/pectin interpolyelectrolyte complex for wound healing application. Advances in Polymer Technology. 2024;**2024**(1):8747902. DOI: 10.1155/2024/8747902

[32] Ferrante M, Álvarez VA, Narain R, Ounkaew A, González JS. Enhancing the properties of chitosan–pectin hydrogels with cellulose nanowhiskers for potential applications in wound dressings. Macromolecular Chemistry and Physics. 2024;**225**(17):2400088. DOI: 10.1002/macp.202400088

[33] Wang H, Tian Z, Wang L, Wang H, Zhang Y, Shi Z. Advancements, functionalization techniques, and multifunctional applications in biomedical and industrial fields of electrospun pectin nanofibers: A review. International Journal of Biological Macromolecules. 2025;**307**:141964. DOI: 10.1016/j.ijbiomac.2025.141964

[34] Martău GA, Mihai M, Vodnar DC. The use of chitosan, alginate, and pectin in the biomedical and food sector—Biocompatibility, bioadhesiveness, and biodegradability. Polymers (Basel). 2019;**11**(11):1837. DOI: 10.3390/polym11111837

[35] Muthulakshmi V, Rajarajeswari GR. In vivo wound healing efficiency of curcumin-incorporated pectin-chitosan biodegradable films. New Journal of Chemistry. 2021;**45**(38):17918-17929. DOI: 10.1039/D1NJ03477D

[36] Mladenovska K et al. Colon-specific delivery of 5-aminosalicylic acid from chitosan-Ca-alginate microparticles. International Journal of Pharmaceutics. 2007;**342**(1):124-136. DOI: 10.1016/j.ijpharm.2007.05.028

[37] Demchenko V, Riabov S, Sinelnikov S, Radchenko O, Kobylinskyi S, Rybalchenko N. Novel approach to synthesis of silver nanoparticles in interpolyelectrolyte complexes based on pectin, chitosan, starch and their derivatives. Carbohydrate Polymers. 2020;**242**:116431. DOI: 10.1016/j.carbpol.2020.116431

[38] Coimbra P et al. Preparation and chemical and biological characterization of a pectin/chitosan polyelectrolyte

complex scaffold for possible bone tissue engineering applications. International Journal of Biological Macromolecules. 2011;**48**(1):112-118. DOI: 10.1016/j.ijbiomac.2010.10.006

[39] Phonrachom O, Charoensuk P, Kiti K, Saichana N, Kakumyan P, Suwantong O. Potential use of propolis-loaded quaternized chitosan/pectin hydrogel films as wound dressings: Preparation, characterization, antibacterial evaluation, and in vitro healing assay. International Journal of Biological Macromolecules. 2023;**241**:124633. DOI: 10.1016/j.ijbiomac.2023.124633

[40] Alvarado N, Abarca RL, Linares-Flores C. Use of chitosan-based polyelectrolyte complexes for its potential application in active food packaging: A review of recent literature. International Journal of Molecular Sciences. 2023;**24**(14):11535. DOI: 10.3390/ijms241411535

[41] Lima M et al. Development of chitosan-based surfaces to prevent single- and dual-species biofilms of *Staphylococcus aureus* and *Pseudomonas aeruginosa*. Molecules. 2021;**26**(14):4378. DOI: 10.3390/molecules26144378

[42] Ghorbani M, Roshangar L, Rad JS. Development of reinforced chitosan/pectin scaffold by using the cellulose nanocrystals as nanofillers: An injectable hydrogel for tissue engineering. European Polymer Journal. 2020;**130**:109697. DOI: 10.1016/j.eurpolymj.2020.109697

[43] Ziani K, Fernández-Pan I, Royo M, Maté JI. Antifungal activity of films and solutions based on chitosan against typical seed fungi. Food Hydrocolloids. 2009;**23**(8):2309-2314. DOI: 10.1016/j.foodhyd.2009.06.005

[44] Hagesaether E, Hiorth M, Sande SA. Mucoadhesion and drug permeability of free mixed films of pectin and chitosan: An in vitro and ex vivo study. European Journal of Pharmaceutics and Biopharmaceutics. 2009;**71**(2):325-331. DOI: 10.1016/j.ejpb.2008.09.002

[45] Espitia PJP, Du W-X, de Jesús Avena-Bustillos R, de Fátima Ferreira Soares N, McHugh TH. Edible films from pectin: Physical-mechanical and antimicrobial properties - a review. Food Hydrocolloids. 2014;**35**:287-296. DOI: 10.1016/j.foodhyd.2013.06.005

[46] Song K et al. Preparation of pectin-chitosan hydrogels based on bioadhesive-design micelle to prompt bacterial infection wound healing. Carbohydrate Polymers. 2023;**300**:120272. DOI: 10.1016/j.carbpol.2022.120272

[47] Birch NP, Barney LE, Pandres E, Peyton SR, Schiffman JD. Thermal-responsive behavior of a cell compatible chitosan/pectin hydrogel. Biomacromolecules. Jun 2015;**16**(6):1837-1843. DOI: 10.1021/acs.biomac.5b00425

[48] Morello G et al. A thermo-sensitive chitosan/pectin hydrogel for long-term tumor spheroid culture. Carbohydrate Polymers. 2021;**274**:118633. DOI: 10.1016/j.carbpol.2021.118633

[49] Tong WY et al. Development of sodium alginate-pectin biodegradable active food packaging film containing cinnamic acid. Chemosphere. 2023;**336**:139212. DOI: 10.1016/j.chemosphere.2023.139212

[50] Seyedebrahimi R, Razavi S, Varshosaz J, Vatankhah E, Kazemi M. Beneficial effects of biodelivery of brain-derived neurotrophic factor and gold nanoparticles from functionalized electrospun PLGA scaffold for nerve tissue engineering. Journal of

Cluster Science. 2021;**32**(3):631-642. DOI: 10.1007/s10876-020-01822-7

[51] Nguyen DH et al. Microencapsulation of curcumin by ionotropic gelation with surfactants: Characterization, release profile and antioxidant activity. Journal of Drug Delivery Science and Technology. 2022;**76**:103812. DOI: 10.1016/j.jddst.2022.103812

[52] Eivazzadeh-Keihan R et al. Recent advances on biomedical applications of pectin-containing biomaterials. International Journal of Biological Macromolecules. 2022;**217**:1-18. DOI: 10.1016/j.ijbiomac.2022.07.016

[53] Sonwane S, Bonde S, Bonde C, Chandarana C. Advances in gelatin-based scaffolds for tissue engineering applications: A review. Journal of Drug Delivery Science and Technology. 2025;**107**:106789. DOI: 10.1016/j.jddst.2025.106789

[54] Freiberg S, Zhu XX. Polymer microspheres for controlled drug release. International Journal of Pharmaceutics. 2004;**282**(1):1-18. DOI: 10.1016/j.ijpharm.2004.04.013

[55] Helmy AM, Elsabahy M, Soliman GM, Mahmoud MA, Ibrahim EA. Development and in vivo evaluation of chitosan beads for the colonic delivery of azathioprine for treatment of inflammatory bowel disease. European Journal of Pharmaceutical Sciences. 2017;**109**:269-279. DOI: 10.1016/j.ejps.2017.08.025

[56] Chourasia MK, Jain SK. Polysaccharides for colon targeted drug delivery. Drug Delivery. 2004;**11**(2):129-148. DOI: 10.1080/10717540490280778

[57] George M, Abraham TE. Polyionic hydrocolloids for the intestinal delivery of protein drugs: Alginate and chitosan — A review. Journal of Controlled Release. 2006;**114**(1):1-14. DOI: 10.1016/j.jconrel.2006.04.017

[58] Szymańska E, Winnicka K. Stability of chitosan—A challenge for pharmaceutical and biomedical applications. Marine Drugs. 2015;**13**(4):1819-1846. DOI: 10.3390/md13041819

[59] Souza VGL et al. Strategies to improve the barrier and mechanical properties of pectin films for food packaging: Comparing nanocomposites with bilayers. Coatings. 2022;**12**(2):108. DOI: 10.3390/coatings12020108

[60] Mishra N et al. Crosslinked and PEGylated pectin chitosan nanoparticles for delivery of phytic acid to colon. International Journal of Pharmaceutics. 2023;**639**:122937. DOI: 10.1016/j.ijpharm.2023.122937

[61] Saberian M, Safari Roudsari R, Haghshenas N, Rousta A, Alizadeh S. How the combination of alginate and chitosan can fabricate a hydrogel with favorable properties for wound healing. Heliyon. 2024;**10**(11):e32040. DOI: 10.1016/j.heliyon.2024.e32040

Chapter 4

Cubosomes; a Promising Carrier for Encapsulation of Drugs, Alginate-Based Compounds and Theranostic Applications

Tanzeela Masood, Sajid Asghar, Ikram Ullah Khan, Muhammad Irfan, Khuriah Abdul Hamid, Hannis Fadzillah Mohsin, Mashani Mohamad, Wan Azman Wan Ismail and Syed Haroon Khalid

Abstract

Cubosomes, a novel class of nanoparticles, have gained substantial interest in the field of drug delivery because of their distinctive bicontinuous cubic structure. Cubosomes are made up of amphiphilic lipid and stabilizers and provide robust and adaptable platform for enclosing a variety of drugs. Because of their natural high surface area and capacity to encapsulate both hydrophilic and hydrophobic materials, cubosomes are incredibly effective in drug delivery applications. Their potential for therapeutic applications is further enhanced by their biocompatibility and sustained drug release. Cubosomes provide intriguing drug delivery system for oral, ocular, dermatological, and intravenous applications. Alginate-based cubosomes represent a significant advancement in targeted drug delivery systems. Their pH-responsive and nontoxic nature, combined with the ability to control the release of encapsulated drugs, makes them a versatile platform for colon-targeted multidrug delivery. Researchers have focused their views on diagnostic aspect of cubosomes. Their hydrophobic volume and minimal surface allow the encapsulation of imaging agent and therapeutic agent in combination. In diagnosis, cubosomes loaded with imaging agent allow for real-time tracking and imaging of medication distribution and therapy response. The integration of therapeutic and diagnostic agent in cubosomes heralds a new era in nanomedicine.

Keywords: nanoparticles, cubosomes, drug deliver, theranostic delivery, glyceryl monooleate

1. Introduction

Cubosomes are nanoparticles that are inverse bicontinuous lyotropic liquid crystals. When amphiphilic molecules self-assemble into a thermodynamically stable structure, they form liquid crystals [1]. Among the various amphiphilic compounds, the hydrophobic effect's ability to cause amphiphilic lipids to self-assemble may result in the formation of some clearly defined, thermodynamically stable structures like lamellar, hexagonal, and bicontinuous cubic phases. These structures are collectively called lyotropic liquid crystals, all of which possess structural symmetry and an average degree of molecule orientation [2]. The dispersion of these structures in water leads to the formation of nanostructures. Liposomes are formed by the dispersion of the lamellar phase, hexosomes are formed by the dispersion of the hexagonal phase, and Cubosomes are formed by the dispersion of the cubic phase in the aqueous phase. These structures have attracted a lot of attention recently due to their distinctive physicochemical characteristics, particularly their potent capacity to encapsulate a wide range of active molecules from hydrophilic, hydrophobic, and amphiphilic classes (**Figure 1**) [4].

The cubic phase is a bicontinuous lipid layer enclosing two water channels but nonintersecting and spread as 3D network-like honeycombed structure. Cubic phases have also many usages in medicinal drug delivery system because of their 3D nanostructure with domains that are hydrophobic and hydrophilic. Despite of all lipid contents are environment friendly and edible; the wide interfacial area can create a tough diffusion canal for the continuously release of drug molecules that have been entrapped. The large size and increase viscosity of the bulk materials pose a challenge for cubic phases in several real-world applications. This raises issues with *in vivo* transport for medication delivery methods like intravenous, oral, and nasal, among other things [5]. Cubic phase is formed by lipid bilayer membrane that separate two distinct nonintersection hydrophilic channel made 3D structure like honeycomb. Characteristics of cubic phase have interfacial area of 400 m^2/g, thickness of 3.5 nm bilayer membrane 3.5 nm, and diameter of pores 5 nm.

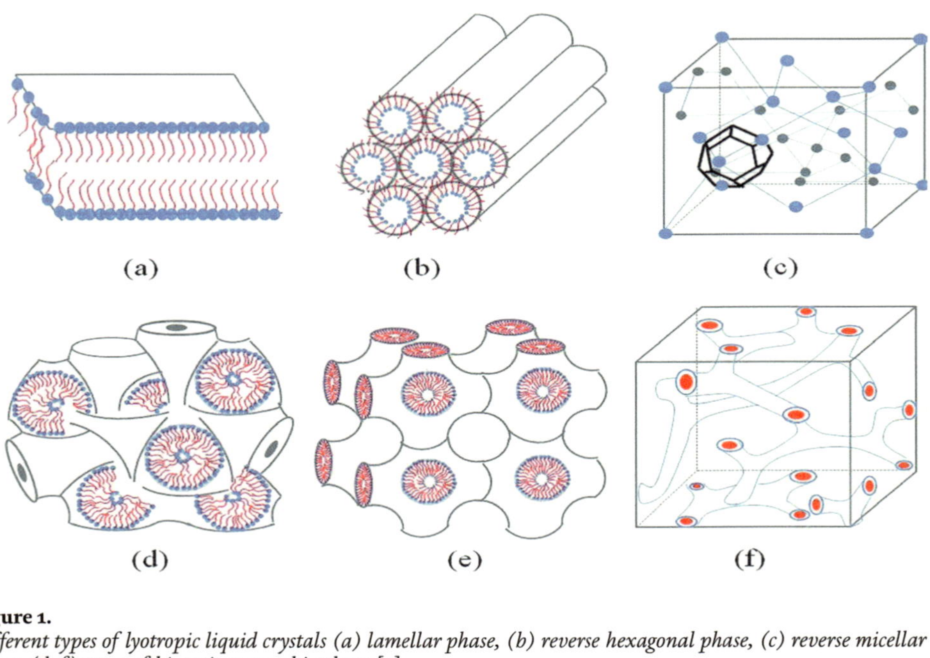

Figure 1.
Different types of lyotropic liquid crystals (a) lamellar phase, (b) reverse hexagonal phase, (c) reverse micellar phase, (d–f) types of bicontinuous cubic phase [3].

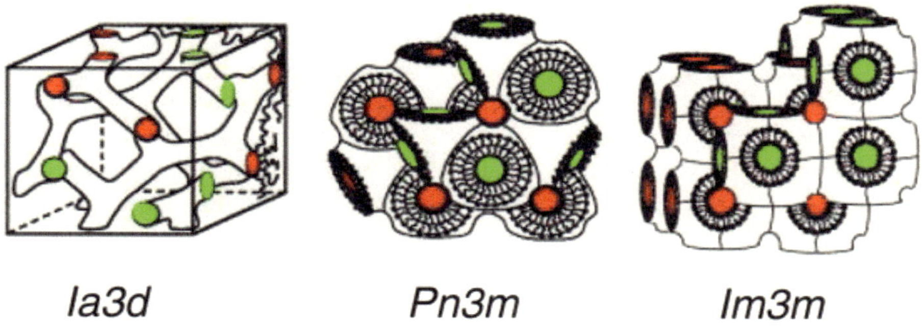

Figure 2.
Symmetries of cubic phase; (lm3m) gyroid space, (Ia3d) diamond space, (Pn3m) primitive space [6].

1.1 Different morphology of cubic phases

i) Primitive surface (Pn3m)

ii) Gyroid surface (Im3m)

iii) Diamond surface (Ia3d)

Cubosomes are lyotropic liquid and bicontinuous lipidic structures that can be found in three several shapes, including gyroid, diamond, and primitive. Their sizes range from 50 to 500 nm. In **Figure 2**, different symmetries of cubic phase are shown that describe three-by-three connections connect each hydrophilic region's channels in the Ia3d diamond structure; four-by-four connections organize each pair of water channels in the Pn3m primitive structure, and six-by-six connections arrange each water channel in the Im3m gyroid structure [7].

2. Cubosomes in drug delivery

Cubosomes are versatile and exciting platform for drug delivery vehicle with several advantages, such as biocompatibility, controlled release, and high loading capacity. Due to their special cubic phase structure and capacity to encapsulate a wide range of medicinal substances, they can be used for various drug delivery applications, such as oral, topical, intravenous, and pulmonary administration. Cubosomes functionality and efficacy are being further developed through ongoing research and development, opening the door for novel and efficient drug delivery methods.

2.1 Oral drug delivery

Cubosomes offer an approach to the many problems associated with the oral delivery of promising medicines, such as high molecular size, poor absorption, and poor aqueous solubility. Large proteins have been encapsulated for local action in the gastrointestinal system as an alternative application. Targeting and controlled release capabilities can be integrated with liquid crystalline nanoparticle technology carriers. The medicine can be distributed effectively *in vivo* due to the particles' ability to develop *in situ* at a regulated rate [8]. Cubosomes have received particular attention for their potential to improve the oral delivery of various substances, including large-molecule medicines and poorly soluble pharmaceuticals [9]. Because of their bioadhesive qualities, ability to interact with the intestinal cell membrane, or ability to secrete physiological surfactants in the gastrointestinal tract (GIT), cubosomes can aid in the absorption of medications taken orally. They can also shield medications from deterioration and have a great loading capacity. Cubosomes are now a viable option for improving the oral administration of poorly soluble medications [9]. The excessive usage of cubosomes in the oral administration of numerous drugs including those that are poorly soluble in water, poorly absorbed, and have large molecular weights has also attracted attention. Due to their bioadhesive qualities, interactions with intestinal cell membranes or increased synthesis of surfactants in the GIT during fat breakdown may aid in the absorption of orally administered medications. Cubosomes are lipid-based nanoparticles that are considered to play important roles in the absorption of drugs. Jin et al. added the anticancer drug protopanaxadiol (PPD) to GMO cubosomes for better oral absorption. The findings showed that the PPD-loaded cubosomes association could raise the Caco cell monolayer model of PPD's permeability values by 53%. Cubosomes effectively encapsulate antioxidant enzyme Q10 that has poor solubility and bioavailability to enhance its oral efficacy and protection against hepaticular damage. Hence, cubosomes provide platform for hepataprotective effect of enzyme and increase its uptake by peripheral tissues [10]. **Table 1** shows the examples of oral drug delivery systems. There are numerous studies that showed the oral application of cubosomes such as clopidogril bisulphate cubosomes to enhance its intestinal release [8].

2.2 Dermatological applications

In transdermal drug delivery, the stratum corneum serves as an enormous obstacle to the skin-penetrating effects of topically applied drugs. On the other hand, because of their unique shape and properties, cubosomes present a potential method for transdermal drug administration. Because of their bioadhesive qualities to the stratum corneum, glyceryl monooleate cubosomes can be employed in various approaches to deliver drugs to the skin and mucous membranes. When used together, microneedles (MNs) and cubosomes have been successfully employed to administer vaccines by the skin. The use of microneedles improved the aqueous mixture's ability to permeate through the skin layers while cumbersome-formulated peptides showed longer skin retention. As a result, it was

Drug	Encapsulation efficiency (%)	Formulation technique	Benefits	References
Ibuprofen	85	Top down(homogenizatiom)	Cubosomes provide sustained property of ibuprofen	[11]
Simvastatin	80	Emulsification technique	Cubosomes enhance the bioavailability of insoluble simvastatin	[12]
Piperine	86.67	Top down(homogenization)	Increase the cognitive function of brain	[13]
Amphotericin b	87.8	Homogenization	Increase bioavailability and reduce nephrotoxicity	[14]

Table 1.
Examples of drugs, encapsulation efficiency, formulation technique, and benefits loaded in cubosomes for oral drug delivery.

discovered that combining MN and cubosomes techniques produced an effective method for delivering antigen locally to the skin cells that were being targeted [15]. Cubosomes have gained attention in melanoma therapy by loading succefully resveratrol that has potential in cancer disease. Resveratrol has limited bioavailability so that it has limited therapeutic activity. But by the virtue of cubosomes nanotechnology, it was possible to enhance its delivery through transdermal drug delivery of cubosomes [16]. Examples of cubosomes used in transdermal drug delivery with their uses are mentioned in **Table 2**.

2.3 Ocular drug delivery

Ophthalmic delivery has many challenges like tear, blinking eyes, and nasolacrimal drainage. The main route for anterior drug absorption, which enables drugs to pass through the ocular tissue, is the cornea. However, the corneal epithelium's lipophilicity and tight conjunction make it the primary barrier to coronal drug absorption, resulting in pulsed drug entrance and inadequate ocular bioavailability of medications [21]. Due to their extended stay at the corneal surface and their adhesive qualities because of the presence of glyceryl monooleate, cubosomes are found to increase the eye bioavailability of the loaded drugs. This is because glyceryl monooleate increases corneal permeability, which in turn increases cubosomes' mucoadhesive properties [22]. In an *in vitro* permeation test, it was discovered that cubosomes preparation raised the apparent permeability coefficient by using cubosomes incorporated with dexamethasone through rabbit eye cornea. Cubosomes preparations are safe and nonirritant for ocular uses. Examples of ocular delivery through cubosomes are shown in **Table 3**. The potential of cubosomes as an ocular drug delivery mechanism for flurbiprofen was investigated by Han et al. By using the Draize method and doing a histological investigation, the cubosomes demonstrated low corneal irritation. The capacity to improve transcorneal permeability was demonstrated by an *in vitro* study of corneal penetration. Flurbiprofen loaded cubosomes had a substantially greater AUC than free eye drops [26].

Drug	Formulation techniques	Encapsulation efficiency (%)	Benefits	References
Capsaicin	Emulsification and homogenization	91–97	Cubosomes provide sustained release and prolong retention time	[17]
Silver sulfadiazine	Emulsification	88	Cubosomes provide an excellent healing effect	[18]
Indomethacin	Top down technique	80	Due to depot effect on the epidermis, cubosomes prolong the anti-inflammatory effect	[16]
Erythromycin	Emulsification	95.29	Cubosomes sustained the effect and treat acne	[19]
Dapsone	Top down	93.7	Increase permeation of dapsone across the epidermal layer and reduce side effects	[20]

Table 2.
Examples of drugs, formulation techniques, encapsulation efficiency, and benefits loaded in cubosomes for dermatological drug delivery.

Drug	Formulation technique	Route of administration	Benefits	References
Dexamethasone	Homogenization	Ocular	Enhancement of retention time and bioavailability	[23]
Ketorolac	Top down	Ocular	Enhance trans corneal permeation and safe for use	[24]
Timolol	Homogenization	Ocular	Decrease intraocular pressure more than commercial product	[25]
Pilocarpine nitrate	Top down	Ocular	Lower ocular irritation and decrease IO pressure	[25]

Table 3.
Examples of drugs, formulation technique, routes of administration, and benefits loaded in cubosomes for ocular drug delivery system.

2.4 Intravenous drug delivery

Leesajakul studied how certain plasma components, including albumin, low-density lipoprotein (LDL), and high-density lipoprotein (HDL), affected the stability and integrity of GMO-based cubosomes as well as how the particles behaved *in vivo* following intravenous injection. Whereas cubosomes bonded with LDL and albumin drive GMO out of the structure, HDL reduced cubosomes to smaller particles containing cubosomes material and HDL [27]. According to Bode et al., cubosomes based on GMO and poloxamar 407 (P 407) had a little but noticeable inclination toward hemolysis and a steric stabilizer (P407) was also ineffective in shielding cubosomes structure from the interactions with blood chemicals. For parenteral usage, poloxamer 188 was recommended rather than poloxamer 407 to create stable cubosomes. Furthermore, another crucial consideration for parenteral administration of the cubosomes dispersions is sterility [28]. Heat treatment transfers the non-cubic particles to cubic particles by autoclaving dispersion for 15 minutes. The existence of larger particles continues to indicate an obstacle for intravenous administration from a pharmacological perspective. To optimize the transformation rate to cubosomes while preventing aggregation, the temperature procedure and dispersion composition must be optimized [29]. A comparison of two nanoparticles delivery systems, the monoolein cubosomes, and nanostructured lipid carriers for the antiparkinson medication bromocriptine was provided by Esposito et al. Bromocriptine was encapsulated with great entrapment effectiveness by cubosomes and lipid carriers. Two behavioral tests (bar test and drag test) were used to assess the impact of bromocriptine loaded nanoparticles on motor impairments in 6-hydroxydopamine hemilesioned rats, a model of Parkinson's disease. Cubosomes showed better results on test than lipid carrier and free drug [30].

3. Alginate-based drug loaded cubosomes

Brown algae naturally contain an anionic polymer called alginate, which is being exploited extensively in pharmaceutical research to create drug delivery systems. Alginate's capacity to create hydrogels by ionic cross-linking with divalent cations, such Ca2+, makes it the perfect material to encase bioactive compounds. Alginate encapsulation technology has gained popularity recently for targeted, topical, and oral medication administration. Alginate-based complexes were used for optimization of curcumin to enhance its bioavailability and stability and to increase the dissolution rate to enhance its anti-inflammatory and anti-oxidant activity [31]. In another study, alginate-loaded microbeads of peppermint oil were prepared to enhance intestinal targeting and increasing the retention time of peppermint oil [32].

Combining the benefits of cubosomes with the biocompatibility and functionality of alginate, alginate-based cubosomes offer a flexible and efficient drug delivery platform [33]. In order to fully exploit their potential in therapeutic applications, more investigation and development are required. Alginate is a naturally derived polysaccharide known for its excellent biocompatibility and biodegradability, making it suitable for biomedical applications. The incorporation of alginate into cubosomes formulations can improve the physical and chemical stability of the cubosomes, protecting the encapsulated drug from degradation. Alginate-based cubosomes offer a controlled release mechanism for drugs, which can be tailored by modifying the alginate content and the crosslinking density. Alginate possesses mucoadhesive properties, which can enhance the residence time of cubosomes at the site of absorption, improving bioavailability. The use of sodium alginate for coating enhances the targeting capability and responsiveness, while the bicontinuous structure of cubosomes facilitates the incorporation of drugs with diverse polarity. These features collectively improve the efficacy and bioavailability of therapeutic agents, demonstrating the potential of alginate-based cubosomes in addressing complex drug delivery challenges and optimizing treatment outcomes [34]. To enhance colon targeting, researchers designed nanocapsules with controlled release cubosomes for pH-triggered delivery of bioactive pharmaceuticals. These capsules were coated with sodium alginate to enable targeted administration, dual controlled release, and increased responsiveness. The bicontinuous structure of cubosomes allows the incorporation of drugs with varying polarity. Thus, researchers successfully developed pH-responsive, nontoxic microcapsules as a versatile platform for colon-targeted multidrug delivery

Alginate-based	Matrix	Applications	References
Polyelectrolyte shell cubosomes	Alginate and chitosan—arginine	pH responsive release of anti-cancer drugs in carcinogenic GIT	[37]
Clindamycin	Ca—alginate and GMO based matrix	Overcome the release issue of clindamycin	[38]
FTIC dextran	MO phase with silk fibroid and alginate	Determination of effect of dextran release with complex	[39]
Colon targeted drug delivery	Microbial nanocellulose cubosomes coated by sodium alginate	pH responsive dual controlled and site specific delivery of bioactive	[35]
Valsartan gel	MO based valsartan gel by using alginate and carbopol, HPMC and guar gum.	Enhancement of bioavailability of valsartan	[36]
Sumatriptan succinate	GMO based cubosomes by using sodium alginate, Kolliphor 188 and samanea saman gum	Brain targeting through nasal route	[40]

Table 4.
Examples of alginate based drugs loaded in cubosomes and applications.

[35]. Additionally, several *in situ* gel formulations of valsartan, an antihypertensive agent, were prepared using sodium alginate and carbopol to enhance absorption [36]. **Table 4** presents the some examples of the drugs loaded in alginate based nano drug delivery systems.

4. Theranostic applications of drug loaded cubosomes

Using nanotechnology in medicine can significantly alter the therapeutic and diagnostic resources used in clinical practice. Nanotechnology provides a special opportunity to advance medication administration and imaging methods [41]. To accomplish this, a wide range of soft and hard matter nanocarriers have been designed and evaluated for their pharmacological, cytotoxic, and physicochemical properties. Liposome compositions, polymeric, gold, iron oxide nanoparticles, dendrimers, and quantum dots are pertinent examples [42]. Initially, the purpose of nanoparticles was to deliver independently active medicinal ingredients or imaging agents to disease locations for either therapeutic or diagnostic purposes. The recent merger of these two elements into a single platform gave rise to the developing field of theranostic nanomedicine, which is the intersection of medicine, chemistry, biology, and pharmacy. Therefore, theranostic nanoparticles can overcome the differences in bio distribution and selectivity of these two separate entities and reduce side effects by actively or passively delivering both medications and imaging probes to the site of disease within the same dosage. Cubosome nanoparticles have been widely used in theranostic applications because of their high surface area and paternal use. Monoolein is prone to lipase enzymes, so to overcome this problem, these nanocubosomes are enclosed by polyethylene glycol coating. This PEG coating shields the nanostructure from opsonization and clearance from the blood stream [43]. Theranostic nanoparticles of cubosomes are also used to enhance the targeting abilities of agents. Researchers have doped the squarine, an NIR emitting fluorescent probe and an anticancer agent, camptothecin with cubosomes because of its internal structure. The internal structure of cubosomes contains nonintersecting water channels and a high surface area that makes cubosomes capable of being utilized in imaging [44]. Cubosomes stabilized by pluronic 108 and folate conjugate reveal more stability and affinity toward the tumor site. Therefore, cubosomes effectively provide the platform for theranostic applications.

Targeted administration of chemotherapeutic agents using cubosomes as a unique drug delivery method has been successfully done with better bioavailability, pharmacokinetics, and safety characteristics of the loaded medications. 5-fluorouracil, a hydrophilic anticancer medication, was delivered *via* a cubosomal drug delivery system for liver targeting. The results showed that the 5-FU-loaded cubosomes considerably enhanced the liver concentration, a value that was near five times more than that found with a 5-FU solution [45]. Reports regarding the successful simultaneous insertion of a squarin based near-infrared (NIR) fluorescent probe and camptothecin have been made. Furthermore, combinations of Pluronic F108 and folate-conjugated Pluronic F108 were used to distribute them in order to demonstrate targeting abilities on cubosomes. The enhanced capacity of these novel nanosystems to target cancer cells was demonstrated through an examination of the lipid droplet changes that HeLa cells experienced after being exposed to manufactured cubosomes [44]. Monophosphoryl lipid A, imiquimod, an immune response modulator, and Toll-like receptor agonists have been introduced to a cubosomes based system. The matrix of cubosomes was discovered to release the model antigen, ovalbumin, in a regulated

Therapeutic agent	Imaging agent	Uses	References
Doxorubicin	Folateprotein	Anticancer	[47]
Etoposide	Rhodamine B	Antiproliferative	[48]
Camptothecin	Squarainprobe	Anticancer	[49]
Quercetin	Dannsylprobe	Anticancer	[50]
Danuorubicin	$NaYF_4$	Anticancer	[44]
Paclitaxel	Lipid	Anticancer	[51]

Table 5.
Theranostic applications of drug loaded cubosomes, imaging agents, and uses.

manner both *in vitro* and *in vivo*. Furthermore, the cubosomes adjuvant combination strongly stimulated CD8+ and CD4+ T cell proliferation, interferon-g, and ovalbumin specific antibody production [46]. Compared to liposome-adjuvant combination and alum, the immunostimulator typically used in vaccines, cubosomes-adjuvant combination was more effective in eliciting OVA-specific cellular and humoral responses. **Table 5** has the theranostic examples of cubosomes and elaborating the therapeutic and diagnostic imaging agents.

In nanomaterial-based cancer theranostic, it is essential to identify biomarkers expressed in malignant cells but not in normal cells. The formulation's nanoparticles also need to be biocompatible, safe, and inactive in a systemic environment. For instance, Zhang and associates made cubosomes using RYLO and a stabilizer known as Poloxamer 407. The cubosomes were coated with poly-ε-lysine to provide long-term medication administration and enhanced effectiveness, and they employed cisplatin and paclitaxel to prevent the drug immediate side effects. Cubosomes drug delivery might sustainably take place using HepG2 cells. Impedance testing and fluorescence imaging were used in cells to assess the therapeutic potential of cubosomes. Cubosomes that were loaded with paclitaxel and functionalized with biotin as the target agent created by Aleandri et al. [52]. A hydrophobic fluorescent lipid was used to identify their cubosomes so that HeLa cells could track their cellular absorption. To transport odoranalectin-targeted cubosomes to the brain, Wu et al. tagged them with the lipophilic fluorescent coumarin-6 to study their brain distribution [53].

5. Conclusions

Bicontinuous cubic phase lyotropic liquid crystals called cubosomes are gaining much attention in drug delivery because of its bio adhesion and biodegradable properties. Translating the concept of cubosomes from drug delivery to theranostic nanoparticles requires the development of nontoxic and biocompatible material such as satbilizers to adjust them according to applications. Cubosomes provide a platform for therapeutic drug delivery especially alginate-based compounds and diagnostic applications. Cubosomes are effectively encapsulated anticancer drugs and florescent agent in combination and represent intriguing delivery for theranostic probes. The potential of cubosomes nanoparticles to enfold different therapeutics and deliver them through different routes makes them crucial strategy in healthcare and diagnostic field. Cubosomes have been shown to significantly improve the efficacy and safety of drug delivery systems as long as research on them continues.

Future recommendation

According to recent research, cubosome nanoparticles have the potential to facilitate prolonged drug release and drug delivery. However, additional optimization is still needed depending on the manner of administration, frequency of dosing, and mode of drug releases. Many researchers have been focusing on the incorporation of protein and peptides and gene vector into cubosomes and more work required to do on these advancements. More studies should be established on the stability of these particles in biological fluids, interactions with cell membranes, cellular absorption mechanisms, intracellular trafficking, control of cell death processes, and reactions connected to infusion. Developing scalable manufacturing processes that ensure the consistency and stability of theranostic cubosomes is essential for their commercial viability. This includes ensuring that the integration of diagnostic and therapeutic agents does not compromise the stability or functionality of the cubosomes. Research should investigate the use of targeting ligands or antibodies conjugated to alginate-based cubosomes for specific disease sites, such as tumors or inflamed tissues. This approach can improve the selectivity and efficiency of drug delivery, reducing off-target effects and enhancing therapeutic outcomes.

Conflict of interest

The authors declare no conflict of interest.

Author details

Tanzeela Masood[1], Sajid Asghar[1], Ikram Ullah Khan[1], Muhammad Irfan[1], Khuriah Abdul Hamid[2,3], Hannis Fadzillah Mohsin[2,3], Mashani Mohamad[2,3], Wan Azman Wan Ismail[2,3] and Syed Haroon Khalid[1,2,3*]

1 Department of Pharmaceutics, Faculty of Pharmaceutical Sciences, Government College University, Faisalabad, Pakistan

2 Faculty of Pharmacy, Universiti Teknologi Mara, Cawangan Selangor, Kampus Puncak Alam, Bandar Puncak Alam, Selangor, Malaysia

3 Innovative Drug Development and Delivery Research Group, Faculty of Pharmacy, Universiti Teknologi Mara, Cawangan Selangor, Kampus Puncak Alam, Bandar Puncak Alam, Selangor, Malaysia

*Address all correspondence to: haroonkhalid@uitm.edu.my; syedharoonkhalid@gcuf.edu.pk

References

[1] Kaur SD et al. Cubosomes as potential nanocarrier for drug delivery: A comprehensive review. Journal of Pharmaceutical Research International. 2021;**33**(31B):118-135

[2] Garti N et al. Lipid polymorphism in lyotropic liquid crystals for triggered release of bioactives. Food and Function. 2012;**3**(7):700-713

[3] Huang Y, Gui S. Factors affecting the structure of lyotropic liquid crystals and the correlation between structure and drug diffusion. RSC Advances. 2018;**8**(13):6978-6987

[4] Chang DP et al. Non-lamellar lipid liquid crystalline structures at interfaces. Advances in Colloid and Interface Science. 2015;**222**:135-147

[5] Ericsson B et al. A cubic protein-monoolein-water phase. Biochimica et Biophysica Acta (BBA)-Biomembranes. 1983;**729**(1):23-27

[6] Rizwan SB, Boyd BJ. Cubosomes: Structure, Preparation and Use as an Antigen Delivery System. Subunit Vaccine Delivery: Springer; 2014. pp. 125-140

[7] Seddon J, Templer R. Polymorphism of lipid-water systems. Handbook of Biological Physics. 1995;**1**:97-160

[8] El-Laithy HM et al. Cubosomes as oral drug delivery systems: A promising approach for enhancing the release of clopidogrel bisulphate in the intestine. Chemical and Pharmaceutical Bulletin. 2018;**66**(12):1165-1173

[9] Kossena GA et al. A novel cubic phase of medium chain lipid origin for the delivery of poorly water soluble drugs. Journal of Controlled Release. 2004;**99**(2):217-229

[10] Mohsen AM et al. Cubosomes as a potential oral drug delivery system for enhancing the hepatoprotective effect of coenzyme Q10. Journal of Pharmaceutical Sciences. 2021;**110**(7):2677-2686

[11] Dian L et al. Cubic phase nanoparticles for sustained release of ibuprofen: Formulation, characterization, and enhanced bioavailability study. International Journal of Nanomedicine. 2013;**8**:845-854

[12] Lai J et al. Glyceryl monooleate/poloxamer 407 cubic nanoparticles as oral drug delivery systems: I. In vitro evaluation and enhanced oral bioavailability of the poorly water-soluble drug simvastatin. AAPS PharmSciTech. 2009;**10**:960-966

[13] Elnaggar YS et al. Novel piperine-loaded tween-integrated monoolein cubosomes as brain-targeted oral nanomedicine in Alzheimer's disease: Pharmaceutical, biological, and toxicological studies. International Journal of Nanomedicine. 2015;**10**:5459-5473

[14] Yang Z et al. Development of amphotericin B-loaded cubosomes through the SolEmuls technology for enhancing the oral bioavailability. AAPS PharmSciTech. 2012;**13**:1483-1491

[15] Kurangi B et al. Formulation and evaluation of resveratrol loaded cubosomal nanoformulation for topical delivery. Current Drug Delivery. 2021;**18**(5):607-619

[16] Esposito E et al. Cubosome dispersions as delivery systems for

percutaneous administration of indomethacin. Pharmaceutical Research. 2005;**22**:2163-2173

[17] Peng X et al. Characterization of cubosomes as a targeted and sustained transdermal delivery system for capsaicin. Drug Design, Development and Therapy. 2015;**9**:4209-4218

[18] Morsi NM et al. Silver sulfadiazine based cubosome hydrogels for topical treatment of burns: Development and in vitro/in vivo characterization. European Journal of Pharmaceutics and Biopharmaceutics. 2014;**86**(2):178-189

[19] Khan S et al. Topical delivery of erythromycin through cubosomes for acne. Pharmaceutical Nanotechnology. 2018;**6**(1):38-47

[20] Nithya R et al. Cubosomes of dapsone enhanced permeation across the skin. Journal of Drug Delivery Science and Technology. 2018;**48**:75-81

[21] Verma P, Ahuja M. Cubic liquid crystalline nanoparticles: Optimization and evaluation for ocular delivery of tropicamide. Drug Delivery. 2016;**23**(8):3043-3054

[22] Maheshwari R et al. Novel application of hydrotropic solubilization in the analysis of some NSAIDs and their solid dosage forms. Indian Journal of Pharmaceutical Sciences. 2007;**69**(1):101-106

[23] Gan L et al. Self-assembled liquid crystalline nanoparticles as a novel ophthalmic delivery system for dexamethasone: Improving preocular retention and ocular bioavailability. International Journal of Pharmaceutics. 2010;**396**(1-2):179-187

[24] Ali Z et al. Fabrication and evaluation of ketorolac loaded cubosome for ocular drug delivery. Journal of Applied Pharmaceutical Science. 2016;**6**(9):204-208

[25] Huang J et al. Ocular cubosome drug delivery system for timolol maleate: Preparation, characterization, cytotoxicity, ex vivo, and in vivo evaluation. AAPS PharmSciTech. 2017;**18**:2919-2926

[26] Han S et al. Novel vehicle based on cubosomes for ophthalmic delivery of flurbiprofen with low irritancy and high bioavailability. Acta Pharmacologica Sinica. 2010;**31**(8):990-998

[27] Leesajakul W et al. Interaction of cubosomes with plasma components resulting in the destabilization of cubosomes in plasma. Colloids and Surfaces B: Biointerfaces. 2004;**34**(4):253-258

[28] Bode J et al. Interaction of dispersed cubic phases with blood components. International Journal of Pharmaceutics. 2013;**448**(1):87-95

[29] Wörle G et al. Influence of composition and preparation parameters on the properties of aqueous monoolein dispersions. International Journal of Pharmaceutics. 2007;**329**(1-2):150-157

[30] Esposito E et al. Nanoparticulate lipid dispersions for bromocriptine delivery: Characterization and in vivo study. European Journal of Pharmaceutics and Biopharmaceutics. 2012;**80**(2):306-314

[31] Azad AK et al. The fabrication of polymer-based curcumin-loaded formulation as a drug delivery system: An updated review from 2017 to the present. Pharmaceutics. 2024;**16**(2):160

[32] Azad AK et al. Electro-hydrodynamic assisted synthesis of lecithin-stabilized

peppermint oil-loaded alginate microbeads for intestinal drug delivery. International Journal of Biological Macromolecules. 2021;**185**:861-875

[33] Lai J et al. Alginate-based encapsulation fabrication technique for drug delivery: An updated review of particle type, formulation technique, pharmaceutical ingredient, and targeted delivery system. Pharmaceutics. 2024;**16**(3):370

[34] Azad AK et al. Encapsulation of black seed oil in alginate beads as a ph-sensitive carrier for intestine-targeted drug delivery: In vitro, in vivo and ex vivo study. Pharmaceutics. 2020;**12**(3):219

[35] Ferreira FV et al. In situ biofabrication of microbial cellulose capsules carrying cubosomes: Toward colon targeted multidrug delivery. ACS Applied Polymer Materials. 2024;**6**(7):3708-3720

[36] Vishnu P et al. Preparation and evaluation of valsartan cubosomes. International Journal of Pharmaceutical Research. 2020;**12**:287

[37] Pimenta BV et al. Interaction of polyelectrolyte-shell cubosomes with serum albumin for triggering drug release in gastrointestinal cancer. Journal of Materials Chemistry B. 2023;**11**(11):2490-2503

[38] Mohamed AI et al. In-vivo evaluation of clindamycin release from glyceryl monooleate-alginate microspheres by NIR spectroscopy. International Journal of Pharmaceutics. 2015;**494**(1):127-135

[39] Mertins O et al. Advances in the design of ph-sensitive cubosome liquid crystalline nanocarriers for drug delivery applications. Nanomaterials. 2020;**10**(5):963

[40] Tekade A, Ghodke P, Patange A, Patil P. Nanostructured cubosomal in situ nasal gel for the treatment of migraine. Journal of Drug Delivery Science and Technology. 2023;**87**:104797

[41] Janib SM et al. Imaging and drug delivery using theranostic nanoparticles. Advanced Drug Delivery Reviews. 2010;**62**(11):1052-1063

[42] Barreto JA et al. Nanomaterials: Applications in cancer imaging and therapy. Advanced Materials. 2011;**23**(12):H18-H40

[43] Gowda BJ et al. The cubosome-based nanoplatforms in cancer therapy: Seeking new paradigms for cancer theranostics. Environmental Research. 2023. 116894 p

[44] Bazylińska U et al. Hybrid theranostic cubosomes for efficient NIR-induced photodynamic therapy. ACS Nano. 2022;**16**(4):5427-5438

[45] Nasr M et al. In vitro and in vivo evaluation of cubosomes containing 5-fluorouracil for liver targeting. Acta Pharmaceutica Sinica B. 2015;**5**(1):79-88

[46] Chircov C, Grumezescu AM. Basics in nanoarchitectonics. Nanoarchitectonics in Biomedicine. Elsevier; 2019. pp. 1-21

[47] Raniolo S et al. Selective targeting and degradation of doxorubicin-loaded folate-functionalized DNA nanocages. Nanomedicine: Nanotechnology, Biology and Medicine. 2018;**14**(4):1181-1190

[48] Tian Y et al. Folic acid-targeted etoposide cubosomes for theranostic application of cancer cell imaging and therapy. Medical Science Monitor: International Medical Journal of Experimental and Clinical Research. 2017;**23**:2426

[49] Caltagirone C et al. Cancer-cell-targeted theranostic cubosomes. Langmuir. 2014;**30**(21):6228-6236

[50] Murgia S et al. Drug-loaded fluorescent cubosomes: Versatile nanoparticles for potential theranostic applications. Langmuir. 2013;**29**(22):6673-6679

[51] Zhang L et al. Theranostic combinatorial drug-loaded coated cubosomes for enhanced targeting and efficacy against cancer cells. Cell Death and Disease. 2020;**11**(1):1

[52] Aleandri S et al. Biotinylated cubosomes: A versatile tool for active targeting and codelivery of paclitaxel and a fluorescein-based lipid dye. Langmuir. 2015;**31**(46):12770-12776

[53] Wu H et al. A novel small odorranalectin-bearing cubosomes: Preparation, brain delivery and pharmacodynamic study on amyloid-β25-35-treated rats following intranasal administration. European Journal of Pharmaceutics and Biopharmaceutics. 2012;**80**(2):368-378

Chapter 5

Smart Hydrogel Film of Bionanocellulose/Alginate/Polyethylene Glycol/Thermochromic Dye (BNC/ALG/PEG/TD) with Thermo-Responsive Behavior

Nurfarahin Zainal, Ku Marsilla Ku Ishak and Yazmin Bustami

Abstract

We aim to explore bionanocellulose/alginate/polyethylene glycol/thermochromic dye (BNC/ALG/PEG/TD) as a thermo-responsive hydrogel film. In this study, BNC was produced using Pichia kudriavzevii USM-YBP2. Then, the morphology of BNC was observed using transmission electron microscope (TEM). The formation of hydrogel film utilized the ionic crosslinking technique by mixing BNC, ALG, PEG, and thermochromic dye (TD) with Ca2+ ions. Then, the hydrogel film was evaluated for its swelling property, thermo-responsive behavior, and biodegradability. BNC exhibits aggregated fiber and clusters of spherical shape. BNC/ALG/PEG/TD formed a gray film, which indicates the successful integration of TD. Interestingly, it shows acceptable stability in wet and dry conditions, suggesting its robustness and stability. However, BNC/ALG/PEG/TD demonstrated low water content (16.6%) as compared to the control probably because of the increment in crosslink density or improvement of polymer-polymer interaction. The hydrogel film changes drastically to colorless at temperatures >30°C proving its thermo-responsive behavior. Interestingly, this hydrogel film also presented a reversible thermochromism property that enables the hydrogel to reverse back to its original color and morphology. However, it shows slow degradation and probably needs more than 2 weeks to fully degrade. In conclusion, BNC/ALG/PEG/TD can be potentially explored as a smart hydrogel film in a wide area of applications.

Keywords: smart hydrogel film, bionanocellulose, alginate, polyethylene glycol, thermo-responsive

1. Introduction

Hydrogels are fascinating materials made up of chains of hydrophilic polymers with a three-dimensional (3D) network [1, 2]. Their biocompatibility and environmental friendliness allow for extensive application and development of new hydrogel technology [2]. Smart hydrogels come into view as a result of ongoing scientific advancements and discoveries [3]. It is in demand due to its wide range of capabilities since it shows response to either physical, chemical, or biological stimuli. Research on hydrogel has changed from relatively straightforward, water-swollen macromolecular networks to hydrogels that may react to environmental stimuli like pH and temperature [4]. Scientists have shown increasing interest in creating temperature-responsive hydrogels that are both biodegradable and biocompatible, largely due to their impressive properties [5]. Thus, hydrogels that show response to temperature were widely studied and actively developed for various applications. In a recent study by Ma et al. [6], they elaborated that stimulus-responsive TEMPO-oxidized cellulose nanofibers/poly(N-isopropylacrylamide) (TOCN/PNIPAM) can be assembled into a flexible sensor and generate sensing signals when driven by temperature changes to achieve real-time feedback. Another study, [7] reported that poly(2-hydroxypropyl acrylate-*co*-acrylamide) (P(HPA-*co*-AM))/hydroxypropyl cellulose (HPC)/lithium chloride (LiCl) exhibits excellent thermo-responsiveness and provide ideas for the next generation of smart multifunctional electronic skin. Also, several studies reported the feasibility of using thermochromic composite hydrogel film for the fabrication of energy-saving smart windows [8–11]. A study by He et al. [8] proposed that a thermochromic composite hydrogel film based on W-doped vanadium dioxide (W-VO2), poly N-isopropyl acrylamide (PNIPAM), and polyacrylamide (PAM) provides an effective energy-saving and multifunctional smart window. Furthermore, the introduction of thermochromic hydrogel as smart window materials is beneficial due to its excellent thermal response, high radiation-blocking efficiency, cost-effectiveness, biocompatibility, and good uniformity [10].

In recent years, nanocellulose has been extensively used to improve the functionality of certain composites. In fact, it has been used widely to form multifunctional composites since it can be gelatinized to form a three-dimensional porous network structure due to a high concentration of hydroxyl groups [6]. There are 3 types of nanocellulose, namely, nanofibrillated cellulose (NFC), cellulose nanocrystals (CNC), and bacterial nanocellulose/bionanocellulose (BNC), depending on their yield methods, processing conditions, and properties [12, 13]. BNC has made an entry into commercial and medical applications upon its extensive utilization. *Gluconacetobacter xylinus*, which was previously known as *Acetobacter xylinum*, is the first bacterial species producing BNC [14]. Since then, the study of different groups of microorganisms for the biosynthesis of BNC has been widely conducted [15]. BNC has been produced by cultivating the respective microbial for a few days in an aqueous culture media containing suitable nutritional parameters which offers an optimum condition [16]. High elasticity and tensile strength are produced by the fibrillated network of the BNC, which results in very good mechanical characteristics [17]. Due to their distinctive physicochemical characteristics, such as hydrophilicity, chirality, biocompatibility, and biodegradability, these nanoparticles are widely regarded as the "green molecule" of the future [18].

In this study, we aim to produce hydrogel film using bionanocellulose (BNC), alginate (ALG), polyethylene glycol (PEG), and thermochromic dye (TD) that can behave as a thermo-responsive hydrogel. The biosynthesis of BNC was conducted using a static fermentation of indigenous yeast, *Pichia kudriavzevii* USM-YBP2, and further evaluated using TEM analysis. The physical morphology, swelling activity, thermo-responsive behavior, and biodegradable properties of BNC/ALG/PEG/TD hydrogel film are reported.

2. Materials and methods

2.1 Materials

Glycerol 99.5% (MW 92.10 g/mol) and ethanol 95% (MW 46.07 g/mol) were purchased from R&M Chemicals (UK). Yeast extract (MW 319.193 g/mol), sodium alginate (MW 120 k–190 k g/mol), polyethylene glycol (6000), disodium phosphate (MW 141.96 g/mol), and calcium chloride dihydrate (MW 147.02 g/mol) were purchased from Sigma Aldrich (USA). The nutrient broth was purchased from Oxoid (UK), peptone (RM006) was purchased from HiMedia Laboratories (India), citric acid monohydrate (MW 210.14 g/mol) was purchased from BDH Laboratory (UK), molasses (MW 3.5×10^5–4.3×10^5 g/mol) was purchased from EMRO, Malaysia, and thermochromic black pigment dye powder (TD) was purchased from Hali Industry (China). *Pichia kudriavzevii* USM-YBP2 glycerol stock was obtained from Lab 207, School of Biological Sciences, Universiti Sains Malaysia, Penang, Malaysia.

2.2 Methods

2.2.1 Production of BNC using Pichia kudriavzevii *USM-YBP2*

The production of BNC using *P. kudriavzevii* USM-YBP2 was adapted from Jasme et al. [19]. Initially, the yeast glycerol stock, *P. kudriavzevii* USM-YBP2, was thawed and activated by adding 1 mL of the glycerol stock into 50 mL of nutrient broth and was agitated at 200 rpm at 30°C for 24 hours. Then, 10 mL of activated yeast was inoculated into 1 L of Hestrin-Schramm broth (HSB) containing 0.5% (w/v) peptone, 0.5% (w/v) yeast extract, 0.27% (w/v) disodium phosphate, 0.115% (w/v) citric acid, and 0.5% (w/v) of 95% (v/v) ethanol. Following the inoculation, the flask was left on a bench for 7 days at room temperature under static fermentation conditions. The formation of pellicles on the air-liquid interphase was monitored. The morphology of *P. kudriavzevii* USM-YBP2 and pellicles was observed under a light microscope. BNC was centrifuged at 7871 x *g* for 10 minutes. The supernatant was discarded, and the remaining cell pellet was washed. This procedure was repeated three times to remove any excess medium. Then, the cell pellet was immersed and vortexed in 30 mL of 1 M NaOH solution for 15 minutes at 60°C to immobilize the yeast cells before it was subjected to centrifugation and washing. The obtained cell pellet was sonicated for 15 minutes and was incubated in a dialysis membrane for at least 3–5 days. Finally, the mixture was centrifuged and the supernatant was discarded. The cell pellet was stored at −40°C for at least 24 hours before the freeze-drying process. The obtained BNC powder was kept at room temperature (RT) until use.

2.2.2 Characterization of BNC using TEM analysis

BNC was analyzed using TEM at the Electron Microscopy Unit, School of Biological Sciences, Universiti Sains Malaysia, Penang, Malaysia. Firstly, a small amount of BNC powder was suspended in distilled water and was sonicated for 10 minutes. Then, the diluted aqueous suspension of the treated BNC was dried for a few minutes on the 400-mesh copper-coated grid. The morphology of BNC was captured using an FEI TECNAI G2 20 S-twin transmission electron microscope (TEM) operating at 200 kV.

2.3 Formulation of BNC/ALG/PEG/TD hydrogel film

To make thermo-responsive hydrogel film (BNC/ALG/PEG/TD), 0.17% (w/v) thermochromic dye (TD) solution was mixed with 4% (w/v) BNC, 8% (w/v) ALG, and 4% (w/v) PEG. Then, the mixture was added into a beaker containing 30 mL of distilled water and stirred until the mixture was completely homogenized. After that, 4 mL of the blending solution was slowly poured onto a Petri dish containing 4% $CaCl_2$ solution, and the hydrogel film was allowed to crosslink for 5 minutes. Subsequently, the formed hydrogel film was washed and rinsed with dH_2O to remove any excess of $CaCl_2$. Finally, the formed BNC/ALG/PEG hydrogel film was air-dried for 12–18 hours at room temperature. The formed hydrogel film was stored at 4°C for further characterization and testing. For control, a hydrogel film without TD solution was prepared and denoted as BNC/ALG/PEG.

2.3.1 Swelling test

Firstly, the hydrogel film was air-dried at room temperature for 48 hours to obtain dried hydrogel. Then, it was immersed in 80 mL of distilled water. The swelling test was conducted for 5 days. The swollen hydrogel was weighed at 24, 48, 72, 96, and 120 hours. The filter paper was used to remove any excess water. The water content (W_r) of the hydrogel films and the equilibrium swelling ratio (eq_{sw}) were calculated using Eqs. (1) and (2) [20].

$$Water\ content\ (Wr) = \frac{W_S - W_d}{W_s} \times 100\% \qquad (1)$$

$$eqsw = \frac{W_S}{W_d} \qquad (2)$$

where W_s denoted the swelling weight and W_d denoted the dry weight of the hydrogel film. The swelling studies were conducted in duplicates.

2.3.2 Testing of thermo-responsive behavior at different temperatures

BNC/ALG/PEG/TD hydrogel films were placed into glass Petri dishes and incubated overnight at 4°C, room temperature (RT), 30°C and 60°C. The thermo-responsive behavior of the respective hydrogel films was observed and recorded. In order to determine its reversible thermochromic property, the hydrogel film was initially heated until 100°C before cooling down by immersing the hydrogel film in dH_2O for 5 days at RT. The color and physical morphology of hydrogel films after heating and cooling conditions were observed and recorded.

2.3.3 Biodegradability test

The biodegradability of hydrogel films was investigated by soil-burial test. The hydrogel film was introduced to a compost bed in a control environment; 30°C, pH 6.0–8.5, and moisture content of 60%. The biodegradation test was taken for 2 weeks. The changes in the hydrogel film were observed and recorded.

2.3.4 Characterization of hydrogel film using SEM analysis

The selected hydrogel films (before and after the degradation process) were analyzed using SEM at the Electron Microscopy Unit, School of Biological Sciences, Universiti Sains Malaysia, Penang, Malaysia. The samples were sputter coated with a thin layer of palladium or gold alloy to improve surface conductivity and tilted 30° for better observation. The surface views of the hydrogel films were taken at a magnification of 100x–10,000x.

2.4 Statistical analysis

A student's T-test was calculated to compare means between two treatment groups. Statistical significance was defined as a $p < 0.05$. The data are presented as the mean ± standard deviation (SD) of multiple replicates.

3. Result and discussion

3.1 Production of *Pichia kudriavzevii* USM-YBP2

Pichia kudriavzevii USM-YBP2 presented an oval or elliptical shape under a light microscope, as seen in **Figure 1**. Furthermore, the majority of the cells become clustered and in proximity with one another as indicative of budding behavior and is aligned with our previous study in Low et al. [18]. The formation of pellicles was observed at the liquid-air interface of the Hestrin-Schramm broth (HSB) and appeared as an interwoven network of fiber. As reported by Lahiri et al. [21], a visible, three-dimensional (3D), and interconnected reticular pellicle is seen at the medium interface during static fermentation of BNC. This physical trait is a strong indication that the obligate aerobic isolates are generating cellulose [22]. A study by Ul-Islam et al. [23] reported that the entrapped carbon dioxide (CO_2) produced by the bacterial metabolism causes the BNC biofilm to float on the top of the medium. It was also speculated that the obligate aerobes, microorganisms embedded within the cellulose matrix, use oxygen for cell growth while utilizing sugars to build cellulosic biofilms or pellicles [24]. According to Campano et al. [25], bacteria gather at the surface and

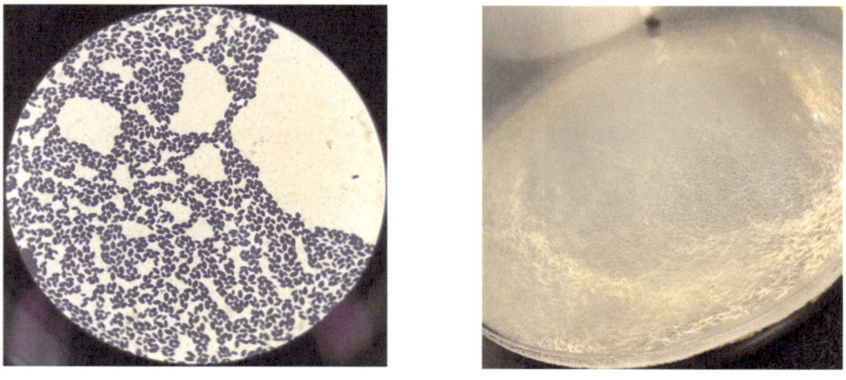

Figure 1.
Morphology observation of P. kudriavzevii *USM-YBP2 under a light microscope with 1000X magnification (left image) and pellicle production on the surface of HSB after 7 days of incubation (right image).*

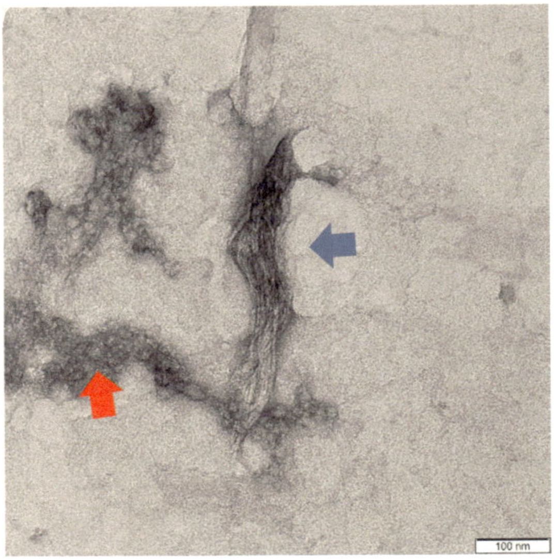

Figure 2.
TEM image of BNC produced using Pichia kudriavzevii *USM-YBP2. A thin and highly aggregated fiber network of BNC (blue arrow) and clustered spherical shapes of BNC (red arrow) were observed under 40,000X magnification.*

produce BNC in layers as the amount of oxygen drops. This is because microorganisms producing cellulose predominantly consume dissolved oxygen during microbial cellulose production.

Transmission electron microscopy (TEM) was used to observe the morphology of BNC produced in static fermentation (**Figure 2**). It can be noticed that BNC shows a densely aggregated fiber network and resembles long needles. Likewise, El-Naggar et al. [26] stated that BNC produced during static culture developed needle-like particles with sizes ranging from 30 to 40 nm. According to Cai and Kim [27], BNC differs from plant cellulose in structure due to a different synthesis procedure in which bacteria synthesize fine sub-elementary cellulose fibrils that are extruded from terminal enzyme complexes into the culture medium during cultivation. Moreover, other types of morphology were also observed. A cluster of spherical shapes BNC was observed in the TEM image and is in agreement with our previous study [18]. Similarly, a study conducted by Zhong et al., Chawla et al., and Trilokesh and Uppuluri [28–30] also reported the formation of spherical CNC (SCNC), and it is a common morphology for BNC produced from fruit sources.

3.2 Formulation and characterization of thermo-responsive hydrogel film

In **Figure 3**, a distinct hydrogel film color can be observed for the control and thermo-responsive hydrogel. The control hydrogel (BNC/ALG/PEG) formed as a translucent sheet of film with a hint of brown color that is due to the addition of BNC. While BNC/ALG/PEG/TD hydrogel formed as a gray film, which indicates the successful integration of TD within the BNC/ALG/PEG hydrogel network. Furthermore, TD was blending well in the BNC/ALG/PEG hydrogel as the gray color was evenly distributed. Interestingly, both hydrogel films show acceptable stability in both wet and dry conditions (**Figure 3A** and **B**). The formed hydrogel films were not only capable of absorbing a large amount of water, but they also could withstand high

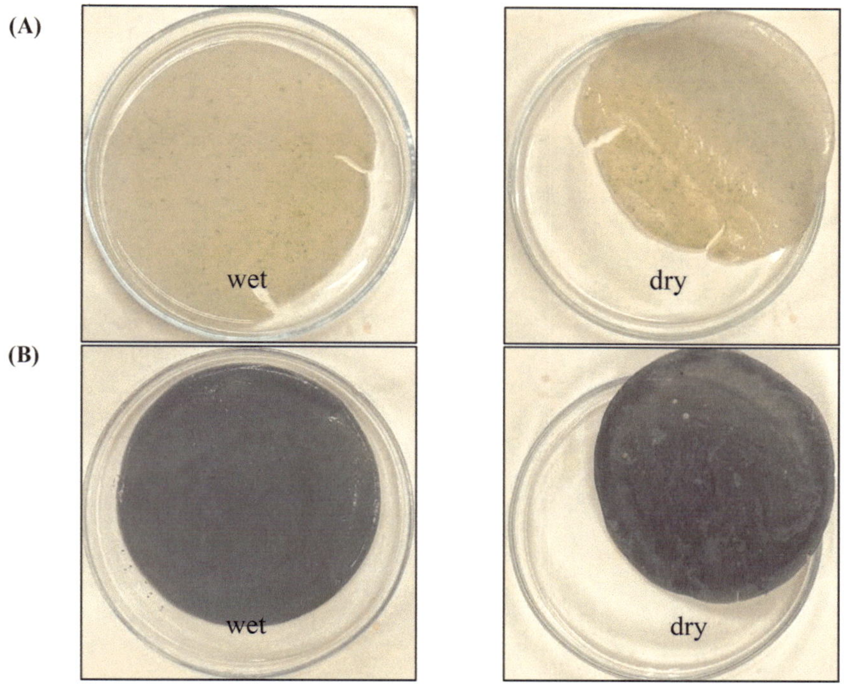

Figure 3.
The wet and dry conditions of (A) BNC/PEG/ALG hydrogel film and (B) BNC/PEG/ALG/TD hydrogel film.

temperatures. This property is important since thermo-responsive hydrogel uses temperature as the external stimulus. Thus, it is necessary to ensure that the hydrogel shows good mechanical strength and is stable at different temperatures, from low to high temperatures. There is no doubt that the integration of BNC helps to improve the mechanical properties and form a robust hydrogel film. A study by Wang et al. [31] reported that nanocellulose-based hydrogels show mechanical stability and create intertwined molecular chains to stabilize rich composite spaces. In a common condition, alginate-hydrogel is not stable and easily disintegrates with the presence of monovalent cations such as Na^+ [32]. Therefore, the addition of BNC is instrumental and helps to improve the mechanical properties of alginate-hydrogel.

Based on **Figure 4**, BNC/ALG/PEG/TD reached swelling equilibrium faster (48 h) than the control, BNC/ALG/PEG (72 h). However, it exhibits significantly low water content (16.6%) as compared to the control (22.5%) ($p < 0.05$). Accordingly, we suggested that the inclusion of TD might affect the hydrogel network as depicted by low water content and equilibrium swelling ratio (1.2). When the crosslink density of the hydrogel increases, swelling decreases, and vice versa [33]. Besides, it is also speculated that the addition of TD might improve the polymer-polymer interaction within the hydrogel network. According to Ghosh and Katiyar [20], the combination of biopolymers has boosted the polymer-polymer contact and further created a hydrogel material with customized features, with polymer-polymer interaction dominating the water-polymer interaction. They also recommended that the polymer-polymer interactions between molecules be improved, which the polymers linked themselves by covalent bonding, hence resulting in disregarding water interactions between the molecules. This property might be causing lower water absorption due to a stronger link between the polymer.

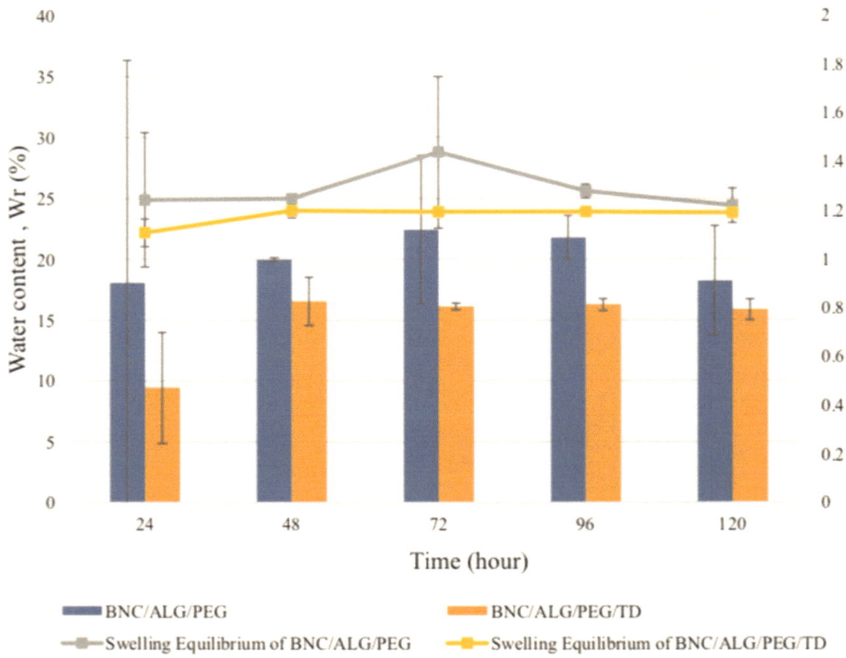

Time (hour)	BNC/ALG/PEG		BNC/ALG/PEG/TD	
	Water content, W_r (%)	Swelling equilibrium ratio, eq_{sw}	Water content, W_r (%)	Swelling equilibrium ratio, eq_{sw}
24	18.10	1.24	9.45	1.11
48	20.00	1.25	16.58	1.20
72	22.45	1.44	16.16	1.20
96	21.81	1.26	16.30	1.20
120	18.25	1.22	15.88	1.19

Figure 4.
Comparison of swelling activity between BNC/ALG/PEG/TD and control, BNC/ALG/PEG hydrogel films depict different water content, W_r*, and swelling equilibrium ratio,* eq_{sw}.

3.3 Thermo-responsive behavior

Figure 5 presents the thermo-responsive behavior of the BNC/ALG/PEG/TD hydrogel film after being treated at different temperatures. It can be seen that at 4°C, the color of the hydrogel film was gray and slightly changed to a lighter gray color at room temperature (RT). Notably, as the temperature increased to 30°C and 60°C, the hydrogel film was colorless and appeared as translucent hydrogel films, similar to the control, BNC/ALG/PEG, in **Figure 3A**. In the present study, BNC/ALG/PEG/TD hydrogel film

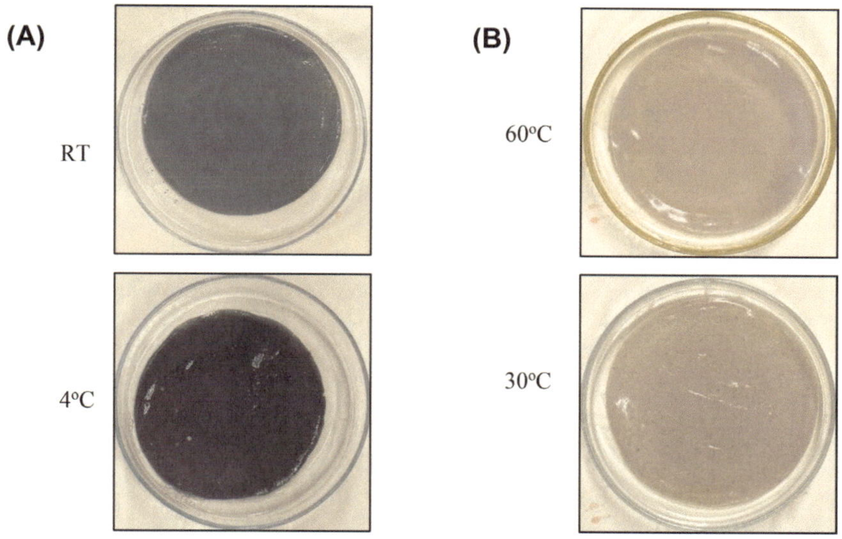

Figure 5.
BNC/ALG/PEG/TD hydrogel film displayed thermo-responsive behavior. At temperatures <30°C; 4°C and RT, hydrogel films appeared as dark and light gray colors, respectively (A). At temperatures >30°C; 30°C and 60°C, hydrogel films appeared as translucent hydrogel films similar to the control, BNC/ALG/PEG.

remains its original gray color at temperatures <30°C, while changing drastically to colorless at temperatures >30°C. This result clearly shows the thermochromism characteristic of the BNC/ALG/PEG/TD hydrogel film in response to the temperature change. A study conducted by Yan et al. [34] elaborated that thermochromic hydrogel film based on photonic nanochains (TRPCF-PNC) exhibited a distinctive and rapid optical change behavior in response to temperatures ranging from 26 to 44°C.

While the BNC/ALG/PEG/TD hydrogel film successfully displayed thermo-responsive behavior, it is important to determine its reversible thermochromic properties. As expected, the hydrogel film was transformed from its original gray color to colorless when heated to 100°C (**Figure 6**). It also can be seen that the film was dehydrated and showed slight deformation, especially in its peripheral area. It is known that hydrogel is rich in water content; thus, exposing hydrogel to high temperatures can cause water evaporation, which contributes to hydrogel dehydration and deformation. Interestingly, the hydrogel film was recovered to its original form when cooling at RT for five consecutive days. The hydrogel film transformed back to a gray color and was rehydrated. Nevertheless, the BNC/ALG/PEG/TD hydrogel film proposed in this study is a promising reversible thermochromic hydrogel, evidenced by its capability to regulate based on the external stimulus, and is found attractive for the fabrication of medical devices, sensors, actuators, soft robotics, and smart windows [6, 11]. The hydrogel actuator is related to its actuated behavior that can absorb and release water. Thus, a thermosensitive hydrogel actuator can be prepared by manipulating the water movement that can be achieved by changing the temperature [6]. Ideally, the solar light getting into the building can be manipulated by the utilization of thermochromic smart windows since it can respond to the atmospheric temperature automatically without additional energy consumption [11]. In the most recent study by Zhong et al. [35], they reported that hydroxypropyl cellulose (HPC) hydrogel smart windows show a significant energy-saving potential compared to ordinary clear glass windows.

Figure 6.
The reversible thermochromic property of BNC/ALG/PEG/TD hydrogel film after heating at 100°C and cooling at RT for five consecutive days.

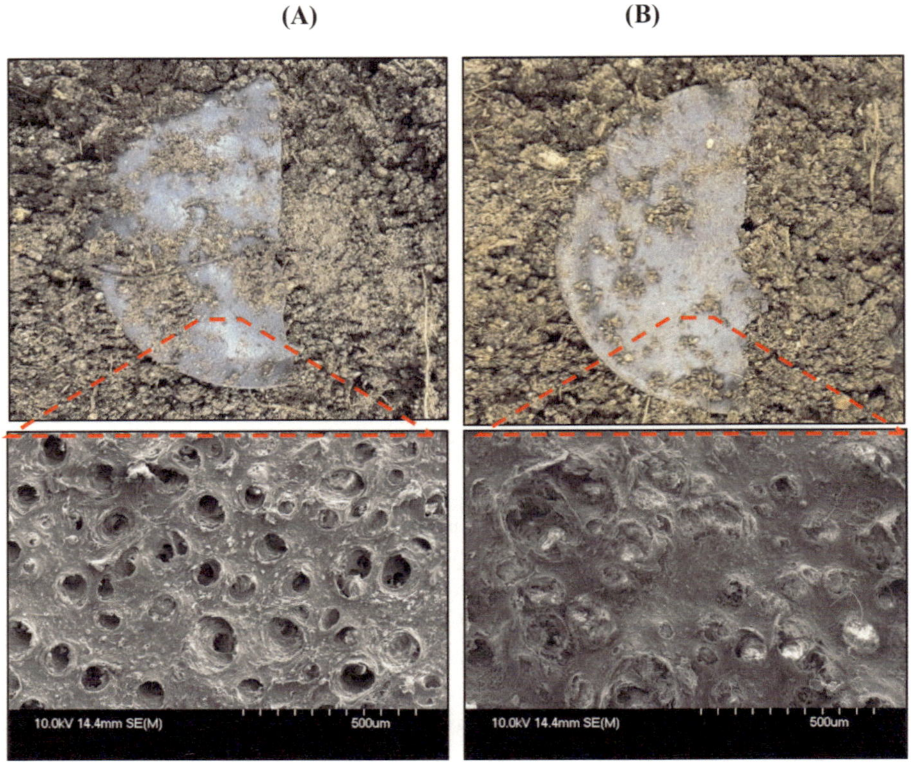

Figure 7.
Observation of the physical morphology and the close-up SEM image of BNC/ALG/PEG/TD hydrogel film (A) before and (B) after 2 weeks of composting.

The biodegradability of BNC/ALG/PEG/TD hydrogel film was investigated under composting environments [moisture content: 60%, 30°C, pH 6.0–8.5]. As presented in **Figure** 7, the physical morphology of the hydrogel film shows minimal damage, with most of the film remaining intact even after 2 weeks of composting. However, when observed under SEM, the porous hydrogel film was not visible, and it might be due to the dissolution of the network chain, and the hydrogel film started to lose its structure. A study by Wang et al. [36] elaborated that hybrid injectable hydrogels consisting of alginate, gelatin, and nanocrystalline cellulose (NCC) have superior mechanical properties resulting from interpenetrated networks with reinforced nanomaterials and good stability against degradation. In fact, alginate-based hydrogel exhibits a remarkably slow degradation rate, as discussed in many studies [36–38]. In the area of bioprinting and tissue engineering applications, alginate shows limitations since mammalian cells do not produce enzymes capable of degrading the polymer, leading to a slow degradation process that is primarily controlled by the outward flux of ions into the surrounding environment [38]. Therefore, it is sufficient to mention that the hydrogel film in this study shows slow degradation under the composting environment, as it probably needs more than 2 weeks to fully degrade and might be useful in certain areas of applications.

4. Conclusions

In conclusion, BNC was successfully produced from *Pichia kudriavzevii* USM-YBP2 by the presence of pellicles at the liquid-air interface of the medium. The morphologies of formed BNC were observed as aggregated fiber and clusters of spherical shape. The addition of BNC is instrumental in improving the mechanical properties of hydrogel film since it shows acceptable stability in wet and dry conditions. The successful integration of thermochromic dye (TD) into the BNC/ALG/PEG network can be easily seen as it formed a gray color film as compared to the control, BNC/ALG/PEG, which formed a translucent film. BNC/ALG/PEG/TD hydrogel film exhibits significantly low water content (16.6%) as compared to the control (22.5%), which is probably because of the increment in crosslink density or improvement of polymer-polymer interaction. Most importantly, BNC/ALG/PEG/TD presented as an excellent thermo-responsive hydrogel as it shows response at different temperatures and displayed a drastic change from gray to colorless when the temperature was >30°C. It also exhibits reversible thermochromic properties as it can be recovered to its original color and morphology when heated to 100°C and cooled at RT. Furthermore, this thermo-responsive hydrogel film shows slow degradation under the composting environment and probably needs more than 2 weeks to fully degrade. Overall, BNC/ALG/PEG/TD is a robust thermo-responsive hydrogel that can be explored in different applications, such as for the fabrication of medical devices, sensors, actuators, soft robotics, and smart windows.

Author details

Nurfarahin Zainal[1], Ku Marsilla Ku Ishak[2] and Yazmin Bustami[1*]

1 School of Biological Sciences, Universiti Sains Malaysia, Gelugor, Penang, Malaysia

2 School of Materials and Mineral Resources Engineering, Universiti Sains Malaysia, Nibong Tebal, Penang, Malaysia

*Address all correspondence to: ybustami@usm.my

References

[1] Duan J-J, Zhang L-N. Robust and smart hydrogels based on natural polymers. Chinese Journal of Polymer Science. 2017;**35**(10):1165-1180. DOI: 10.1007/s10118-017-1983-9

[2] Zainal SH, Mohd NH, Suhaili N, Anuar FH, Lazim AM, Othaman R. Preparation of cellulose-based hydrogel: A review. Journal of Materials Research and Technology. 2021;**10**:935-952. DOI: 10.1016/j.jmrt.2020.12.012

[3] Sikdar P, Uddin MM, Dip TM, Islam S, Hoque MS, Dhar AK, et al. Recent advances in the synthesis of smart hydrogels. Materials Advances. 2021;**2**(14):4532-4573. DOI: 10.1039/d1ma00193k

[4] Buwalda SJ, Boere KWM, Dijkstra PJ, Feijen J, Vermonden T, Hennink WE. Hydrogels in a historical perspective: From simple networks to smart materials. Journal of Controlled Release. 2014;**190**:254-273. DOI: 10.1016/j.jconrel.2014.03.052

[5] Gao S, Tang G, Hua D, Xiong R, Han J, Jiang S, et al. Stimuli-responsive bio-based polymeric systems and their applications. Journal of Materials Chemistry. B, Materials for Biology and Medicine. 2019;**7**(5):709-729. DOI: 10.1039/c8tb02491j

[6] Ma Y, Lu Y, Yue Y, He S, Jiang S, Mei C, et al. Nanocellulose-mediated bilayer hydrogel actuators with thermo-responsive, shape memory and self-sensing performances. Carbohydrate Polymers. 2024;**335**(122067):122067. DOI: 10.1016/j.carbpol.2024.122067

[7] Zhou B, Yuan W. Tunable thermoresponsive and stretchable hydrogel sensor based on hydroxypropyl cellulose for human motion/health detection, visual signal transmission and information encryption. Carbohydrate Polymers. 2024;**343**(122497):122497. DOI: 10.1016/j.carbpol.2024.122497

[8] He Q, Wang Y, Zhou Y, He M, Xu R, Hu S, et al. PAM-PNIPAM/W-doped VO2 thermochromic hydrogel film with high solar modulation capability for smart windows deployment. Optical Materials (Amst). 2019;**97**(109367):109367. DOI: 10.1016/j.optmat.2019.109367

[9] Li M, Wang H, Hu J, Hu J, Zhang S, Yang Z, et al. Smart hydrogels with antibacterial properties built from all natural building blocks. Chemistry of Materials. 2019;**31**(18):7678-7685. DOI: 10.1021/acs.chemmater.9b02547

[10] Chen F, Wu X, Lu G, Nie J, Zhu X. Thermochromic hydrogels with adjustable transition behavior for smart windows. ACS Applied Materials & Interfaces. 2024;**16**(16):21013-21023. DOI: 10.1021/acsami.3c19272

[11] Xu G, Lu Y, Zhou X, Moloto N, Liu J, Kure-Chu SZ, et al. Thermochromic hydrogel-based energy efficient smart windows: Fabrication, mechanisms, and advancements. Materials Horizons. 2024;**11**:4867-4884. DOI:10.1039/D4MH00903G

[12] Nasir M, Hashim R, Sulaiman O, Asim M. Nanocellulose: Preparation methods and applications. In: Cellulose-reinforced Nanofibre Composites. Woodhead Publishing, ScienceDirect; 2017. pp. 261-276

[13] Sharma A, Thakur M, Bhattacharya M, Mandal T, Goswami S. Commercial application of cellulose nano-composites-a review. Biotechnology Reports. 2019;**21**:e00316. DOI: 10.1016/j.btre.2019.e00316

[14] Blanco Parte FG, Santoso SP, Chou C-C, Verma V, Wang H-T, Ismadji S, et al. Current progress on the production, modification, and applications of bacterial cellulose. Critical Reviews in Biotechnology. 2020; **40**(3):397-414. DOI: 10.1080/07388551.2020.1713721

[15] Charoenrak S, Charumanee S, Sirisa-Ard P, Bovonsombut S, Kumdhitiahutsawakul L, Kiatkarun S, et al. Nanobacterial cellulose from kombucha fermentation as a potential protective carrier of lactobacillus plantarum under simulated gastrointestinal tract conditions. Polymers (Basel). 2023;**15**(6):1356. DOI: 10.3390/polym15061356

[16] Thomas B, Raj MC, Joy J, Moores A, Drisko GL, Sanchez C. Nanocellulose, a versatile green platform: From biosources to materials and their applications. Chemical Reviews. 2018; **118**(24):11575-11625. DOI: 10.1021/acs.chemrev.7b00627

[17] Ludwicka K, Kaczmarek M, Białkowska A. Bacterial nanocellulose-a biobased polymer for active and intelligent food packaging applications: Recent advances and developments. Polymers (Basel). 2020;**12**(10):2209. DOI: 10.3390/polym12102209

[18] Low DYS, Supramaniam J, Soottitantawat A, Charinpanitkul T, Tanthapanichakoon W, Tan KW, et al. Recent developments in nanocellulose-reinforced rubber matrix composites: A review. Polymers (Basel). 2021;**13**(4): 550. DOI: 10.3390/polym13040550

[19] Jasme N, Elangovan J, Mohd Yahya AR, Md Noh NA, Bustami Y. First report of biocellulose production by an indigenous yeast, *Pichia kudriavzevii* USM-YBP2. Green Processing and Synthesis. 2022;**11**(1):458-477. DOI: 10.1515/gps-2022-0023

[20] Ghosh T, Katiyar V. Cellulose-based hydrogel films for food packaging. In: Polymers and Polymeric Composites: A Reference Series. Cham: Springer International Publishing; 2018. pp. 1-25

[21] Lahiri D, Nag M, Dutta B, Dey A, Sarkar T, Pati S, et al. Bacterial cellulose: Production, characterization, and application as antimicrobial agent. International Journal of Molecular Sciences. 2021;**22**(23):12984. DOI: 10.3390/ijms222312984

[22] Jozala AF, de Lencastre-Novaes LC, Lopes AM, de Carvalho S-EV, Mazzola PG, Pessoa-Jr A, et al. Bacterial nanocellulose production and application: A 10-year overview. Applied Microbiology and Biotechnology. 2016; **100**(5):2063-2072. DOI: 10.1007/s00253-015-7243-4

[23] Ul-Islam M, Khan S, Ullah MW, Park JK. Bacterial cellulose composites: Synthetic strategies and multiple applications in bio-medical and electro-conductive fields. Biotechnology Journal. 2015;**10**(12):1847-1861. DOI: 10.1002/biot.201500106

[24] Subbiahdoss G, Osmen S, Reimhult E. Cellulosic biofilm formation of Komagataeibacter in kombucha at oil-water interfaces. Biofilms. 2022; **4**(100071):100071. DOI: 10.1016/j.bioflm.2022.100071

[25] Campano C, Balea A, Blanco A, Negro C. Enhancement of the fermentation process and properties of bacterial cellulose: A review. Cellulose. 2016;**23**(1):57-91. DOI: 10.1007/s10570-015-0802-0

[26] El-Naggar NE-A, Mohammed ABA, El-Malkey SE. Bacterial nanocellulose

production using cantaloupe juice, statistical optimization and characterization. Scientific Reports. 2023;**13**(1):51. DOI: 10.1038/s41598-022-26642-9

[27] Cai Z, Kim J. Bacterial cellulose/poly (ethylene glycol) composite: Characterization and first evaluation of biocompatibility. Cellulose. 2010;**17**(1): 83-91. DOI: 10.1007/s10570-009-9362-5

[28] Zhong C, Zhang G-C, Liu M, Zheng X-T, Han P-P, Jia S-R. Metabolic flux analysis of Gluconacetobacter xylinus for bacterial cellulose production. Applied Microbiology and Biotechnology. 2013;**97**(14):6189-6199. DOI: 10.1007/s00253-013-4908-8

[29] Chawla PR, Bajaj IB, Survase SA, Singhal RS. Microbial cellulose: Fermentative production and applications. Food Technology Biotechnology. 2009;**47**(2):107-124

[30] Trilokesh C, Uppuluri KB. Isolation and characterization of cellulose nanocrystals from jackfruit peel. Scientific Reports. 2019;**9**(1):16709. DOI: 10.1038/s41598-019-53412-x

[31] Wang C, Bai J, Tian P, Xie R, Duan Z, Lv Q, et al. The application status of nanoscale cellulose-based hydrogels in tissue engineering and regenerative biomedicine. Frontiers in Bioengineering and Biotechnology. 2021; **9**:732513. DOI: 10.3389/fbioe.2021.732513

[32] Hoffman AS. Hydrogels for biomedical applications. Advanced Drug Delivery Reviews. 2012;**64**:18-23. DOI: 10.1016/j.addr.2012.09.010

[33] Cates RS. Influence of Crosslink Density on Swelling and Conformation of Surface-Constrained Poly(N-Isopropylacrylamide) Hydrogels. USF Tampa Graduate Theses and Dissertations. University of South Florida; 2010

[34] Yan H, Si L, Li G, Zhao L, Luo W, Ma H, et al. Heterogeneous thermochromic hydrogel film based on photonic nanochains. Nanomaterials (Basel). 2022;**12**(11):1867. DOI: 10.3390/nano12111867

[35] Zhong S, Xue Y, Wang K, Wang L, Jiang T. A sustainable utilization approach of waste biomass resources to smart materials for buildings. Materials Today Communications. 2024; **40**(109506):109506. DOI: 10.1016/j.mtcomm.2024.109506

[36] Wang K, Nune KC, Misra RDK. The functional response of alginate-gelatin-nanocrystalline cellulose injectable hydrogels toward delivery of cells and bioactive molecules. Acta Biomaterialia. 2016;**36**:143-151. DOI: 10.1016/j.actbio.2016.03.016

[37] Jeon O, Alt DS, Ahmed SM, Alsberg E. The effect of oxidation on the degradation of photocrosslinkable alginate hydrogels. Biomaterials. 2012; **33**(13):3503-3514. DOI: 10.1016/j.biomaterials.2012.01.041

[38] Barceló X, Eichholz KF, Garcia O, Kelly DJ. Tuning the degradation rate of alginate-based bioinks for bioprinting functional cartilage tissue. Biomedicine. 2022;**10**(7):1621. DOI: 10.3390/biomedicines10071621

Chapter 6

Laboratory and Semi-Industrial Studies on the Application of Chitosan for Efficient Wastewater Treatment

Dilyana Zvezdova

Abstract

The limited and progressively diminishing natural water resources necessitate ongoing research and technological development to identify optimal technical solutions for localized treatment of drinking and wastewater. Local water purification systems are essential for providing additional treatment of drinking water, thereby safeguarding human health in the event of accidental contamination due to leaks, corrosion, deposits, or other issues within the water transport infrastructure leading to the consumer. The need for designing compact water purification systems is clearly substantiated. These systems—including process type and choice of water treatment materials—are intended to perform functions such as removal of mechanical impurities, organic contaminants, and more. This section presents both global and personal experiences in developing innovative solutions for medium- and small-scale localized water treatment, focusing on the use of chitosan in producing filtration, sorption, ion exchange, and other types of media for these compact purification systems. These systems will gradually contribute to practical applications in daily life, demonstrating their global relevance and environmental importance. Installing compact water purification systems for on-site wastewater treatment will highlight the concept of personal environmental responsibility. The study found that chitosan-based materials, when incorporated into the composition of composite ion-exchange sorbents, contribute to a reduction in the percentage of pollutants in wastewater. These materials outperform conventional options, such as activated carbon, in terms of cost-effectiveness and environmental impact. In this context, new technological and structural laboratory and semi-industrial facilities have been developed, offering an alternative and environmentally significant solution for the future treatment of domestic and industrial wastewater.

Keywords: chitosan, small water treatment device, local drinking water treatment, deionization, sorption, ion exchange

1. Introduction

A significant part of domestic consumers and many industrial sectors (pharmaceutical, textile, paper, and plastic manufacturing) consume substantial quantities of water and reagents during the use, production, and coloring of their products. As a result, they generate considerable amounts of contaminated wastewater. These toxic effluents are a primary source of water pollution, causing severe ecological damage and destruction of aquatic flora in the receiving water bodies if not treated using appropriate technological methods. The characteristics of these pollutants include high biological oxygen demand (BOD), chemical oxygen demand (COD), insoluble suspended solids, unpleasant odors, toxicity, dyes, and other contaminants [1–4].

Coloring is the first and most easily recognizable visual and undesirable pollutant in wastewater [5, 6]. Over the past several decades, several methods for the treatment of wastewater streams have been developed and applied. These technologies can be categorized into conventional methods and methods aimed at the extraction and reuse of the recovered components. The literature provides numerous studies on the treatment of colored wastewater streams using these approaches [4].

In practice, no single process can provide comprehensive water treatment. Therefore, it is often necessary to combine multiple processes to achieve the desired quality of purified water in the most economically feasible way. This underscores the need of developing new water treatment methods that are both effective and applicable on local and industrial scales.

In the scientific literature, it is well known that the use of inexpensive sorbents is an effective and economical method for water purification. A wide variety of sorption materials have been presented, and their ability to remove dyes has been extensively studied [2, 3, 7]. However, it should be noted that inexpensive sorbents with high sorption capacities are currently under development in order to reduce the dosage of sorbent in the technological process. Recently, significant attention has been given to various biosorption materials such as biopolymers, fungal biomass, or bacterial biomass, which can be produced in large quantities and are environmentally friendly.

Applied scientific research clearly highlights the need for alternative sorbent materials to replace activated carbon and ion-exchange resins, driven by environmental and economic considerations. This trend arises from the single-use nature of activated carbon in adsorption processes and its environmental impact during its composting (or disposal). Additionally, the production technologies of ion-exchange resins, along with the time and cost associated with reagent-based regeneration, present significant financial challenges.

For these reasons, competitive sorbents such as chitosan and its derivative compounds are being sought to reduce and overcome environmental impacts by contributing to more sustainable and environmentally friendly water purification solutions in the context of global environmental challenges (e.g., freshwater scarcity, industrial pollution).

Special emphasis has been placed on a natural amino-polymer with a polysaccharide nature, such as chitosan.

Chitosan has been extensively studied by researchers as a biosorbent for binding and removing dyes from aqueous solutions. Its advantage as an effective biosorbent is due to its cost-effectiveness compared to commercially available and frequently used activated carbon, as well as its chemical affinity for chelate formation, through the formation of chemical bonds with contaminants and heavy metal ions.

The aim of the present monographic section is to establish the design of small-scale water treatment systems and to investigate their technological behavior in terms of the potential application of these biopolymer materials for water treatment, both for wastewater and natural waters. One of the applied aspects is the final process for decolorizing wastewater. The influence of various parameters, such as the characteristics of chitosan, activation conditions, the diversity of processes, and the experimental conditions used in "batch" systems, on the biosorption process is discussed. The potential applications of chitosan biosorbent as a final process for softening and deionizing natural waters are also referred to, with the aim of achieving the required parameters concerning the content of colloidal and molecularly dispersed substances, as well as residual free chlorine used for disinfection in drinking water supply systems.

2. Chitosan as a sorption material

The majority of polymeric sorbents and ion exchange resins available on the market are petroleum-based products derived from raw materials that, when used in the chemical industry, have environmentally negative impacts. Recently, there has been increasing interest in the development and application of natural, inexpensive alternatives to synthetic polymeric sorbents and ion exchange resins, derived from natural and economically and environmentally sustainable sorbents [7].

As it is well known, chitin is commonly found in the exoskeletons of insects, the cell walls of fungi, and marine organisms, and it is a widely distributed amino polysaccharide in nature. It has also been noted that this inexpensive natural material is a linear homopolymer composed of β-(1-4)-linked N-acetylglucosamine (**Figure 1**).

Chitin is structurally similar to cellulose and is an aminopolymer containing acetamide groups located at the C-2 position, replacing the hydroxyl groups. The presence of these groups offers a significant advantage, providing distinctive, characteristic sorption functions, which also lead to possibilities for modifying interactions. As a raw material, the biopolymer is primarily available on the market through extraction from marine animals with hard coverings [8]. Chitin is extracted from the shells of shrimp, crabs, and fish through acid treatment to dissolve calcium carbonate, followed by alkaline extraction to remove proteins, and then undergoes bleaching to obtain colorless products [9–11].

Since the biodegradation of chitin in the shells of marine waste organisms is a very slow process, the collection of large quantities of waste has become increasingly important, leading to the need for recycling these materials to produce chitin, or chitosan. Their use as sorption or other water treatment materials for wastewater

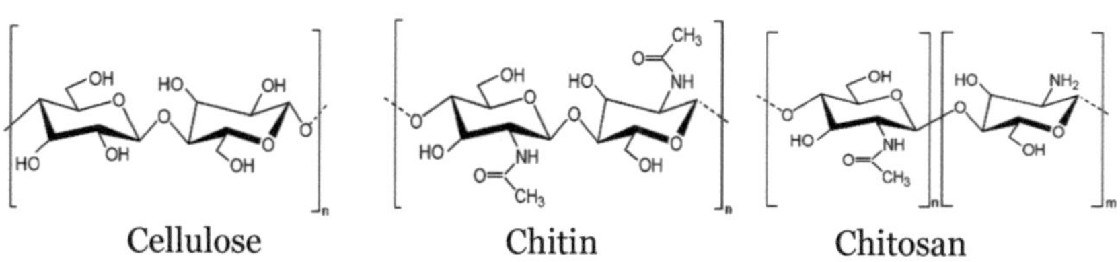

Figure 1.
Chemical structure of cellulose chitin and chitosan.

represents another industrial application, which is not only environmentally significant for wastewater from various industries but also economically viable.

Chitosan is obtained through N-deacetylation of chitin in a strongly alkaline environment, where its macromolecules mainly contain 2-amino-2-deoxy-β-D-glucopyranosyl residues (D-glucosamine – $GlcNH_2$) and, to a lesser extent, 2-acetamido-2-deoxy-β-D-glucopyranosyl residues (N-acetyl-D-glucosamine – GlcNAc) (**Figure 1**). It is of commercial and technological interest due to its higher nitrogen content compared to cellulose. Chitosan is soluble in acidic solutions and is chemically more reactive than chitin and cellulose. The main reason for this is its exceptional ability to attract various substances to its surface and form different chemical bonds, as reflected in numerous publications [9, 12].

Confirming its positive characteristics are its cationic nature, biodegradability, biocompatibility, ability to form film-like surfaces, bioadhesion, multifunctionality, hydrophilicity, and sorption properties [8]. These diverse advantages (**Table 1**) contribute to its classification as a promising raw material for sorption processes, widely used in the purification of natural and wastewater, serving as coagulants [13], flocculants, and chelating agents [14].

In general, an ideal sorbent for dye molecules should meet the following criteria: low cost, availability, sorption capacity, high sorption rate, high selectivity at varying concentrations, and effectiveness in removing a wide range of contaminants from the water to be treated.

However, it should be noted that in practice, it is difficult to find a sorbent that meets all of the requirements for a specific technological case. Therefore, there is a significant need to search for and develop new sorption materials and systems. Additionally, it should be emphasized that the sorption process is a promising technology for water treatment [15], especially if the sorbent has selective properties for removing anionic, cationic, or non-ionic dyes.

The effectiveness of chitin and chitosan for sorbing dye molecules dissolved in aqueous solutions has been reported and published by many authors. They conducted research experiments and found that chitosan and its modified derivatives, with high affinity and selectivity, could remove significant amounts of dye substances from aqueous solutions [16, 17].

Physical and chemical properties	Poly-electrolyte properties	Biological properties
Linear aminopolysaccharide with high nitrogen content, stable D-glucosamine structure, high crystallinity, hydrophilicity, capacity to form intermolecular hydrogen bonds, high viscosity, weak alkalinity, protonated amine groups exhibit pronounced nucleophilicity (pKa = 6.3), insoluble in water, soluble in dilute acidic solutions, contains reactive groups for chemical activation and crosslinking, forms salts with organic and inorganic acids, chelating and complex-forming properties, ion conductivity.	Cationic biopolymer with high charge density; flocculant agent; interacts with negatively charged molecules; sorptive, filtration, and coagulation properties; film-forming properties and adhesiveness; material for isolating biomolecules.	• Biocompatibility: Non-toxic, biodegradable. • Bioactivity: Antimicrobial activity (against fungi, bacteria, and viruses), anti-acid, anti-tumor, and anti-coagulant properties: exhibits hypoepidemic activity, bioadhesion.

Table 1.
Properties of chitosan.

The production of sorption products is a key element that needs to be clarified in order to introduce and define the necessary properties that determine the expansion of its potential applicability as a starting sorption raw material. Chitosan has three types of reactive functional groups: one amine group and two hydroxyl groups—primary and secondary—at positions C-2, C-3, and C-6, respectively (**Figure 1**). Its advantage over other polysaccharides is that its chemical structure allows for specific modifications with minimal difficulty, especially at the C-2 position [9]. These functional groups enable substitution reactions, leading to chemical transformations and the production of a variety of highly useful materials for different targeted applications. The simplest and most easily performed chemical modification and activation is through carboxymethylation, acetylation, and grafting. The resulting derivatives of chitosan can generally be obtained and classified as modified polymers, crosslinked, and composite materials.

One important class of products are the crosslinked chitosan materials, such as gel spherical granules (micro-particles, microspheres, and nanoparticles). Gel products are typically classified into three categories, depending on the nature of their crosslinking, namely: entangled networks, covalent crosslinking, and networks formed by physical interactions. Berger et al. [18] proposed the following modified classification for chitosan gels, dividing them into chemical and physical gels. Hydrogels and spherical granules can self-form as covalently crosslinked polymers. In this type of chemical crosslinking reaction, crosslinking agents are molecules with reactive functional groups that allow the formation of bridges between polymer chains. To date, the most commonly used crosslinking agents for chitosan are: glyoxal; formaldehyde and glutaraldehyde (GA); epichlorohydrin (ECH); ethylene glycol diglycidyl ether (DGDE); and others [12, 18].

In general, the crosslinking procedure is necessary and aims to improve the mechanical strength of the substance and its resistance to chemical environments, particularly in acidic solutions. It also modifies the hydrophobicity and ensures stability under larger pH fluctuations—properties that are crucial for defining a good sorbent. However, this process significantly reduces the number of available free amino groups in the polymeric structure, and as a result, the ligand density and reactivity of the biopolymer are decreased. This also reduces the accessibility and uptake of sorbable substances within the material, leading to a loss of flexibility in the polymer chains due to crosslinking. This explains the reduction in diffusion access within the inter-contact volumes of the porosity of the sorption granules. This chemical procedure can lead to a significant decrease in dye removal efficiency and sorption capacity. Therefore, it is important to understand, characterize, and control the conditions during crosslinking interactions, as these determine and allow modification of the crosslinking density, which is the primary parameter influencing practical properties of the gel sorbent [18].

The choice of a specific method from the available options depends on factors such as thermal and chemical stability and the requirements regarding particle size. Modification of chitosan through covalent interactions allows for the reduction of the influence of sorbent particle size [19]. These changes are expected to be caused by the magnitude of the specific surface area and the fractional particle size, which influences the diffusion mechanism. The equilibrium time increases with the increase in crosslinking agent concentration [20].

In practice, methods are often combined and include various necessary sequential treatments [21]. The emulsion crosslinking method is widely used for synthesizing microspheres. With this method, particle sizes can be controlled by altering the size of the water droplets. Another interesting method is spray-drying and countercurrent droplet movement in a turbulent drying stream environment, where temperature and humidity of the environment are varied. Chitosan microspheres obtained by this technique are characterized by high sphericality and large specific surface area. Along with the formation of complex formations with negatively charged ions or molecules, an interesting property of chitosan is its ability, as a gel, to interact with specific polyanions. These gel-forming processes occur due to the formation of intermolecular and intramolecular crosslinking, where the polyanions serve as the crosslinking medium. Sodium tripolyphosphate (TPP) is most commonly used for ionotropic gelation of chitosan during constant stirring to form porous structures of granules (flakes or powder) with distinct inter-contact porosity [12, 22].

Chiou et al. [22] found that crosslinking as part of the conditioning process for the chitosan sorbent is necessary to increase mechanical strength and improve the material's stability in acidic and alkaline environments, which also enhances the sorbent's sorption technological lifespan. The sorption capacity increases significantly at low degrees of substitution and decreases as substitution increases [20, 23]. Gordana et al. [24] published a review of methods for preparing magnetite-chitosan granules and data regarding their application for wastewater treatment in "batch" processes (i.e., under static conditions), as well as suitable methods for loading them into filtration water treatment systems through dynamic filtration sorption processes.

At the end of the technological lifespan of biosorbents and their disposal, they degrade in the natural environment into harmless substances such as carbon dioxide, water, and other environmentally safe, non-toxic products.

3. Experiments under static and dynamic conditions

The establishment of a sorption system of liquid/solid phase requires two types of studies: sorption tests under static conditions and investigations in dynamic, continuous flow systems. This section discusses research data regarding sorption under static conditions, as well as some technical and technological solutions for water treatment facilities for ion exchange and final sorption under dynamic filtration conditions.

Research under static conditions is based on the fact that the sorption phenomenon, which occurs at the liquid-solid interface, leads to a change in the concentration of the solution. Experimental processes are simple and easy to implement, which makes them frequently preferred for application in medium-sized and small technological systems, where sorption components (sorbent and sorbate) are mixed in suitable reservoirs. The advantages of the static method lie in the fact that parameters such as sorbate concentration, sorbent quantity, pH, ionic strength, temperature, and so on are easily controllable.

4. Technological design and structural solutions for water treatment facilities using biopolymeric, synthetic polymer, and ion exchange materials

Drinking water sources are constantly influenced by processes such as diffusion, point-specific interactions, morphology, and specificity. The trend toward the appearance of local water treatment systems on the market is becoming increasingly relevant, reflecting the demand for new, innovative design and technological solutions.

A specific area of research in this development is the structural essence of water treatment facilities and their functional effectiveness when purifying water from the public drinking water supply system, with corresponding local applications for domestic drinking, laboratory, and/or small-scale production needs. The main focus of the research is on exploring the potential for using single-body, multi-chamber, multiprocess water treatment systems, localized for the further purification of water for the aforementioned needs.

This combination aims to create technologies and systems for multi-stage biological, filtration, ion-exchange, and sorptive water treatment, with the objective of simultaneously performing final filtration, sorption, and ion-exchange treatments on the biologically purified water stream, ultimately returning it to a state suitable for reuse [15].

5. Project-technology and experimental research for the creation and technological behavior of small-scale water treatment systems for local drinking water treatment

The process of creating small-scale water treatment facilities for processing water from the public water supply system includes the following ordered project-technological and experimental investigations:

- Preparation of sorption material and implantation of elementary silver
- Implantation of elementary silver into the physical porous structure or onto the sorption surface of biopolymer sorbents like chitosan, synthetic polymer sorbents, activated carbon, and so on
- Microscopic observations of the porous structure
- Observations of the porous structure of non-implanted and silver-implanted sorbents
- Preparation and investigation of sorbent-ion exchange resin mixtures for natural water treatment
- Preparation and examination of the behavior of a mixture of implanted sorbent and ion exchange resin when purifying natural water
- Investigation of sorption properties of sorbent-ion resin mixture
- Laboratory experiments to test the sorption properties of the sorbent-ion resin mixture using model water
- Investigation of properties of magnetite filtration fillers with chitosan additives
- Study of magnetite filtration fillers with an alternative chitosan coagulant for contact-coagulation treatment of natural water to remove colloidal and dispersed impurities.

- Design and technological scheme for small-scale water treatment systems
- Development of technical designs and technological schemes for local water treatment systems aimed at various technological alternatives:
  - o Design of a two-chamber system with a conical body.
  - o Design of a three-chamber multi-stage (multiprocess) system with either a cylindrical or conical body.
  - o Technological project calculations for the manufacturing of water treatment facilities and creating a technological scheme for connection to the drinking water supply system.
- Experimental investigations on operational parameters and technological behavior of the systems
- Experimental studies to determine the operational parameters and technological behavior of the designed water treatment facilities.
- Comparative evaluation of key technological elements and parameters
- Comparative evaluation of the main technological elements and parameters, such as: type of construction, flow velocity, type of fillers, sorption and/or ion exchange capacity, purification efficiency of the created systems
- Project-technology and experimental research for the creation and evaluation of small-scale deionization units for drinking water
- Research on deionizers for producing water in accordance with international standards such as: BDS ISO 3696-93 "Water for Laboratory Analysis"; ASTM D 1193–99; DIN 38404–52.

5.1 Preparation of sorbent material (activated carbon, granular chitosan)

For the experimental studies, granular chitosan with a heterogeneous granulometric composition has been chosen as the filling material for the small-scale water treatment system. The sorbent is prepared by sieving through a set of sieves with sizes: 0.3, 0.6, 0.8, and 1.2 mm to obtain a defined granulometric composition. The prepared sorption material is then subjected to further conditioning procedures to ensure its functionality for the intended application.

5.2 Implantation of silver into the physical porous structure of granular chitosan

The prepared sorbent undergoes special treatment to apply elementary silver onto its sorption surface, that is, into the physical porous structure that forms the "skeleton" of the sorbent. The procedure is carried out as follows:

- Samples of chitosan are placed in three glass columns and left in contact with a 5% $AgNO_3$ solution for different times (t): 30, 45, and 60 minutes.

- The $AgNO_3$ solutions are drained from the columns at a linear flow rate of 2–4 m/h, and the same volumes of the solution are recirculated by passing it four times through the packed filling at the aforementioned linear flow rates.

- The fillings in the columns are washed with two relative volumes of deionized water at flow rates of 2–4 m/h.

- A final wash of the fillings in the columns is performed by passing deionized water (over 6 relative volumes) until a negative test result is achieved, indicating the absence of silver ions in the waste washout streams.

- The packed granular chitosan in the columns is then evacuated of any residual moisture-laden particles by vacuuming with a laboratory vacuum pump for 15 minutes.

Microscopic observations of the cross-sections of the grains from the samples before and after the $AgNO_3$ treatment are shown in **Figures 2** and **3**, illustrating the structural changes in the chitosan granules before and after silver implantation.

5.3 Preparation of a mixture of sorbent (activated carbon, chitosan, and ion exchanger) suitable for drinking water purification

For the simultaneous implementation of the process of local sorptive and ion-exchange post-treatment of drinking water, a mixture has been prepared consisting

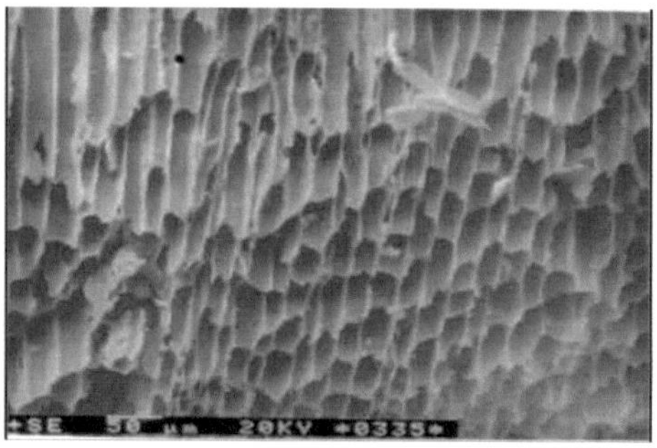

Figure 2.
Scanning electron microscopy (SEM) image of a sample before treatment with $AgNO_3$.

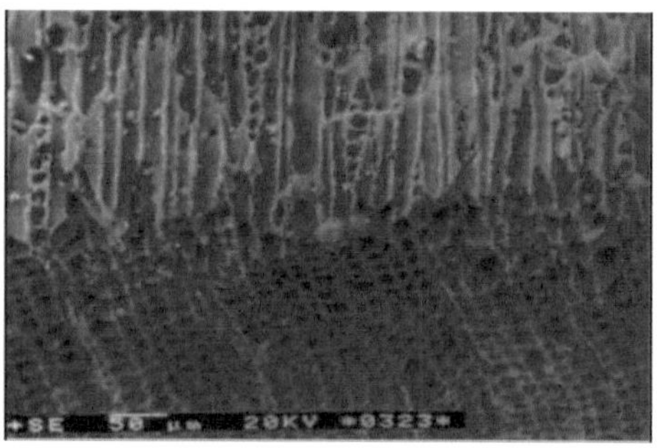

Figure 3.
Scanning electron microscopy (SEM) image of a sample after treatment with $AgNO_3$.

of: a mixture of conditioned sorbent (activated carbon, chitosan) with a weakly acidic cation exchanger (carboxyl type) in hydrogen ion form, a mixture of sorbent (activated carbon and chitosan) with a strongly acidic cation exchanger in sodium ion form, and a strongly basic anion exchanger in chloride ion form.

The essence of the idea for creating the mixture is to add a certain amount of weakly acidic cation exchanger in hydrogen ion form to the sorbent, which, upon contact with the treated water, exchanges hardness cations (Ca^{2+}; Mg^{2+}) from bicarbonate compounds, that is, the so-called temporary hardness. Activated carbon or chitosan with silver implanted in its structure prevents the formation and growth of microorganisms throughout the entire operational period of the filtration media. Through the sorption process, the sorbent also removes harmful impurities from the water, such as organic humic substances, ammonia halogen derivatives, iron and manganese-containing compounds, active chlorine, and others.

For the preparation of the mixed sorptive-ion exchange media described above, it is important to investigate the porous structure of the components (**Figures 2–4**), as well as the conditioning of the weakly acidic, strongly acidic, and strongly basic ion exchangers to convert them into the aforementioned ionic forms (i.e., H-form, Na-form, and Cl-form) according to established methodologies [25].

With the two pre-conditioned components (sorbent and cation exchanger) present in a 2:1 ratio for their joint functionality, the next step is their homogeneous mixing through mechanical stirring. Based on preliminary predictive information regarding the qualitative and quantitative characteristics of the water to be post-treated, this ratio should be adjusted as follows:

- For waters with very high total hardness, the proportion of the cation exchanger component should be increased.
- For water with high concentrations of contaminants above the maximum permissible concentration (MPC), which can be removed via the sorption process, the proportion of the sorbent should be increased.

The ionites from the second sorption–ionite mixture, used for the removal of nitrates from water, were not subjected to structural-porosimetric studies, as they

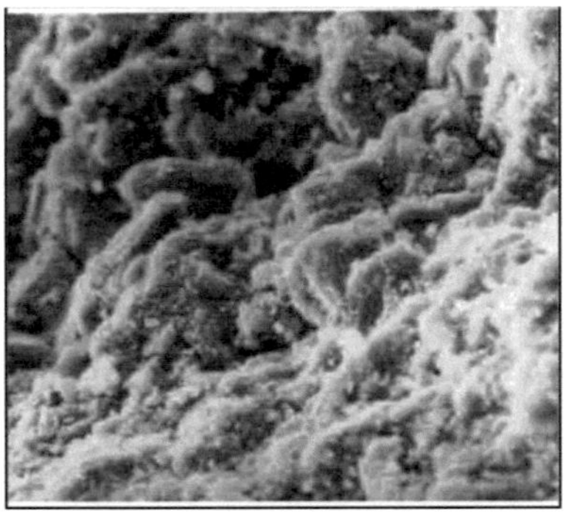

Figure 4.
Structural porosimetry results: Electron microscope photograph of the polyreticular weakly acidic cation exchanger Dowex CCR-2 after the working period.

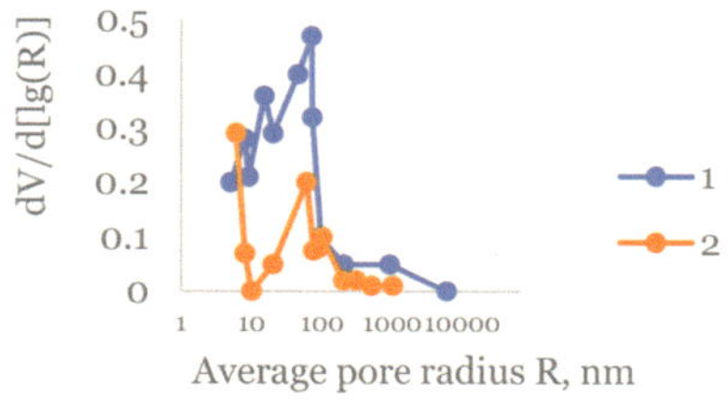

Figure 5.
Differential pore size curves of the polyreticular weakly acidic cation exchanger Dowex CCR-2: Curve 1 – before the working period; curve 2 – after the working period.

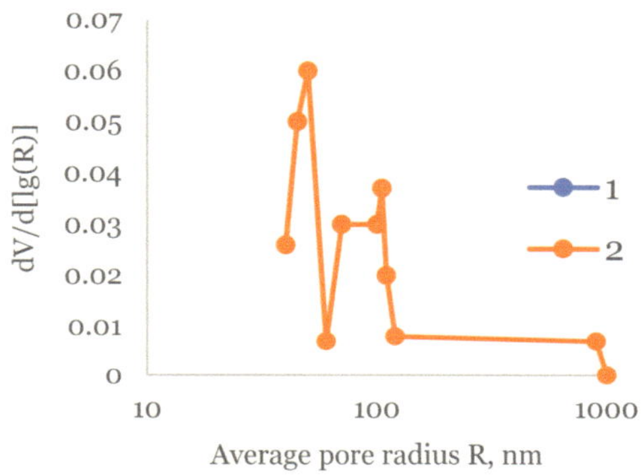

Figure 6.
Differential pore size distribution curves of granulated chitosan: Curve 1 – before the operating period; curve 2 – after the operating period.

are of the so-called monoreticular [15] (gel) structural type, and their "pseudo-porosity," in terms of dimensionality, is beyond the minimum boundary range of possible usable methods.

From the structural-porosimetric results shown in **Figures 5** and **6**, it can be observed that the physical porosity of the components in the water purification mixture changes. The pore size distribution curves indicate a narrowing of the pore radius due to their blockage with colloidal impurities (organic humus substances, etc.) sorbed during the operational cycle, a process known as fouling.

The ionites from the second sorption–ionite mixture, used for the removal of nitrates from water, were not subjected to structural-porosimetric studies, as they are of the so-called monoreticular (gel) structural type, and their "pseudo-porosity," in terms of dimensionality, is beyond the minimum boundary range of possible usable methods [15].

6. Studies of the filtration-purification properties of sorbent-ion-exchange compositions through laboratory and pilot experiments

After preparing the mixture of the components mentioned above, the sorption-ion-exchange properties of the sorbent–ion-exchange mixture are studied through laboratory experiments with model water. To carry out the experiment, model water was prepared with approximately 0.600 mg/dm^3 of iron and manganese and 55–75 mg/dm^3 of nitrates. The exact concentrations were measured for each newly prepared volume of model water using an analytical method from the commonly accepted standardized methods (BDS 3425–85, OH-9, Drinking Water, 1986; BDS 15149–80, Drinking Water, 1981) suitable for the respective purification. The model water is filtered through the filler made of the appropriately prepared homogeneous mixture, which is packed into a column with a diameter of D = 20 mm, cross-sectional area F = 0.0314 dm^2, and height H = 1000 mm (layer height Hsl = 580–600 mm). The filtration experiment is carried out by passing the model water with a flow rate of Qavg = 3 dm^3/h, or a linear velocity W = 7.9 m/h. The total volume of the homogeneous mixture filler in the column is V = 0.190 dm^3. During the experiment, samples are taken from the filtrate at every 51 relative volumes. Analyzes were conducted for the content of: Fe, Mn; organic substances (permanganate oxidizability); turbidity; pH; specific electrical conductivity; total hardness; chlorides; alkalinity according to methyl orange. The results obtained from the laboratory experiments with the mixture, including activated carbon, chitosan with implanted silver, and weakly acidic cationite Dowex CCR-2, are reflected in **Table 2**.

Through the process combination studied by us, laboratory and pilot experiments were conducted with model natural waters having different concentrations of salts. The synthetic polymeric ion-exchangers used in these experiments were Dowex HCR-S and Dowex SBR with a monoreticular (gel) physical structure.

The ion exchangers were conditioned using conventional, most commonly used methods [26]. Their reagent regeneration was carried out with a 10% sodium chloride solution at a specific regenerant consumption of 160 g/dm^3. The specific load during the operating period was 12 dm^3/dm^3.h.

The model water solutions had varying contents of: nitrates, sulfates, and bicarbonates (Am.o). The experiments were conducted with the model waters listed in **Table 3** using the ion-exchange pilot installation shown in **Figure 7**. The ion-exchange modules F1 and F2 were loaded with strong acid cationite and strong base anionite, respectively, converted into sodium and chloride ion forms. The results from the laboratory experiments with a mixture that includes activated carbon with implanted silver or chitosan, strong acid cationite in sodium ion form Dowex HCR-S, and strong base anionite Dowex SBR in chloride form are shown in **Tables 3** and **4**.

The technological experience in water treatment shows that very often the functionality of synthetic polymeric, biopolymeric sorption and ion-exchange materials used for the treatment of natural waters is compromised due to the presence of natural organic colloidal dispersed impurities (humic substances) in them [27, 28].

The most commonly applied technological approach for eliminating their harmful impact is the coagulation treatment of natural waters using the volume coagulation method, with the dosing of appropriate coagulant doses [28]. However, the efficiency of these processes largely depends on the type and dose of the chosen coagulant. They are very inconvenient for use in small-scale technological devices for local water purification. This work investigates the possibility of implementing a reagent-free contact-coagulation process by filtering natural water through a mechanical filter loaded with magnetite black sand with an appropriate grain size composition, that is, the so-called "Siroflock" process in dynamic filtration conditions. The design of the technological system also allows for the dosing of chitosan as a coagulant/flocculant.

Constructive technical solutions for mechanical filters have been sought, which are most suitable for the contact-coagulation process and the magnetite filler loaded into the filtration vessel (the so-called "immedium" filters). Some publications and

Ratio volume number	Measured parameters for the raw and treated water - filtrate											
	Turbidity, mg/dm^3		P. O. mgO$_2$/dm^3		Ho meq/dm^3		Specific el condactivity µS/cm		Mn^{2+} mg/dm^3		Fe total mg/dm^3	
	R. W. *	F*	R. W.	F*	R. W.	F*	R. W.	F*	R. W.	F*	R. W.	F**
51	1, 8	0, 48	2, 39	0, 57	3, 4	1, 71	284	296	0, 60	0, 03	0, 54	0, 11
102	1, 6	0, 47	2, 39	0, 54	3, 4	1, 69	282	291	0, 60	0, 04	0, 57	0, 10
153	1, 5	0, 39	2, 20	0, 57	3, 4	1, 68	281	300	0, 62	0, 04	0, 56	0, 09
204	1, 6	0, 39	2, 39	0, 53	3, 4	1, 69	283	294	0, 63	0, 05	0, 51	0, 10
255	1, 4	0, 38	2, 20	0, 55	3, 4	1, 65	285	296	0, 62	0, 05	0, 55	0, 12
306	1, 7	0, 34	2, 21	0, 56	3, 4	1, 70	283	298	0, 65	0, 06	0, 58	0, 13
357	1, 6	0, 31	1, 90	0, 58	3, 4	1, 67	286	292	0, 61	0, 06	0, 52	0, 13
408	1, 5	0, 29	1, 84	0, 59	3, 4	1, 69	284	297	0, 64	0, 07	0, 53	0, 15
459	1, 6	0, 30	2, 45	0, 61	3, 4	1, 68	282	300	0, 61	0, 07	0, 57	0, 14
510	1, 5	0, 26	2, 45	0, 63	3, 4	1, 75	281	270	0, 62	0, 08	0, 55	0, 16
561	1, 4	0, 26	2, 31	0, 59	3, 4	1, 91	285	286	0, 58	0, 08	0, 52	0, 15
612	1, 3	0, 24	1, 95	0, 58	3, 4	1, 98	284	286	0, 57	0, 08	0, 56	0, 19
663	1, 5	0, 24	2, 42	0, 60	3, 4	2, 7	285	285	0, 58	0, 09	0, 51	0, 18
714	1, 6	0, 22	1, 98	0, 62	3, 4	2, 8	281	281	0, 55	0, 14	0, 52	0, 20
765	1, 4	0, 22	2, 44	0, 61	3, 4	2, 8	283	285	0, 57	0, 18	0, 51	0, 23

**R.W – raw untreated model water.*
***F – treated model water (filtrate).*

Table 2.
Experimental results from the filtration experiment with filler in a column with D = 20 mm; filler height H = 600 mm; filler volume V = 190 cm^3; flow rate Q = 3 dm^3/h.

№	Parameter	Raw water	№1	№2	№3	№4	№5
1	turbidity mg/dm^3	1,6	0,85(46%)	0,3 (81%)	0,3 (81%)	0,2 (86%)	0,4 (62%)
2		1,4	0,2 (86%)	0,3 (79%)	0,2 (86%)	0,2 (86%)	0,3 (79%)
3		1,4	0,2 (86%)	0,2 (86%)	0,2 (86%)	0,2 (86%)	0,3 (79%)
1	Hardness m eq/dm^3	3,2	1,4 (56%)	1,7 (47%)	3,2 (0%)	3,2 (0%)	1,5 (53%)
2		3,2	1,8 (44%)	1,8 (44%)	3,2 (0%)	3,2 (0%)	1,7 (47%)
3		3,2	2,0 (38%)	2,2 (31%)	3,2 (0%)	3,2 (0%)	1,9 (41%)
1	Mn^{2+} µg/dm^3	49	26 (47%)	3,0 (94%)	5,5 (89%)	2,6 (95%)	2,2 (95%)
2		40	27 (32%)	1,6 (96%)	4,5 (88%)	2,5 (94%)	2,5 (94%)
3		43	32 (26%)	2,2 (95%)	5,2 (88%)	2,2 (95%)	2,4 (94%)
1	Fe total µg/dm^3	181	54 (70%)	36 (80%)	51 (72%)	40 (78%)	36 (80%)
2		154	22 (86%)	26 (83%)	40 (74%)	31 (80%)	29 (81%)
3		152	30 (80%)	30 (80%)	46 (70%)	35 (77%)	33 (78%)
1	P. O. mgO$_2$/dm^3	2,1	0,6 (71%)	0,4 (81%)	0,3 (86%)	0,1 (95%)	0,1 (95%)
2		2,0	0,5 (75%)	0,3 (85%)	0,1 (95%)	0,1 (95%)	0,1 (95%)
3		1,9	0,5 (74%)	0,5 (74%)	0,1 (95%)	0,1 (95%)	0,1 (95%)

№	Parameter	Raw water	№1	№2	№3	№4	№5
1	Cl^- m eq/dm^3	0,2	0,2 (0%)	0,2 (0%)	1,8	2,1	0,2 (0%)
2		0,3	0,3 (0%)	0,3 (0%)	1,6	2,3	0,3 (0%)
3		0,2	0,2 (0%)	0,2 (0%)	1,5	2,2	0,2 (0%)
1	pH	7,6	7,6 (0%)	7,6 (0%)	7,6 (0%)	7,6 (0%)	7,6 (0%)
2		7,8	7,8 (0%)	7,8 (0%)	7,8 (0%)	7,8 (0%)	7,8 (0%)
3		7,5	7,5 (0%)	7,5 (0%)	7,5 (0%)	7,5 (0%)	7,5 (0%)
1	Ам.о. m eq/dm^3	1,9	0,6 (68%)	0,4 (37%)	0,3 (84%)	П О М	1,9 (0%)
2		2,0	0,6 (70%)	0,6 (35%)	0,7 (65%)	П О М	2,0 (0%)
3		2,0	0,6 (70%)	1,0 (30%)	0,7 (65%)	П О М	2,0 (0%)

Note: In **Table 3**, *the columns labeled from №1 to №5 represent the following: №1 – a filling of weakly acidic cationic resin Dowex CCR-2 and activated carbon, conditioned in the laboratory; №2 – a filler from the company "Brita" (Germany); №3 – a filling of strongly basic anion resin "Duolite A101D" in chloride ion form and chitosan (sample after the first stage); №4 – a filling of strongly basic anion resin "Duolite A101D" in chloride ion form and activated carbon (sample after the second stage); №5 – a filling from a factory module purchased from the company "Deplan;" B.D.M – below the detectable minimum.*

Table 3.
Experimental results from laboratory experiments of the studied fillings for comparison (as a degree of reduction [%]) for small-scale water treatment systems.

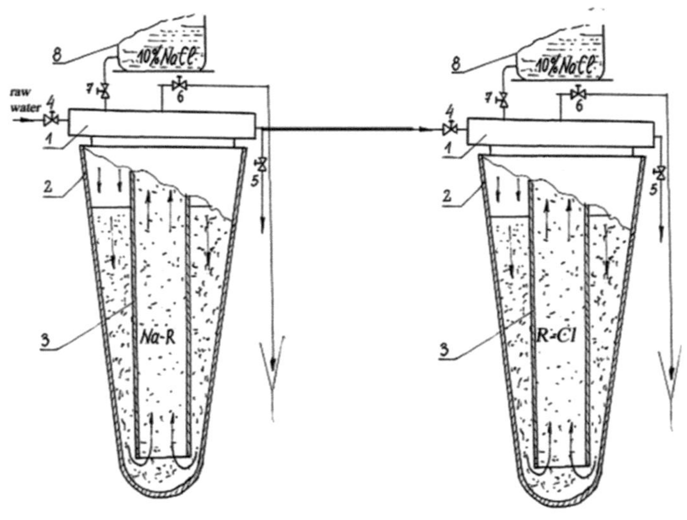

Figure 7.
Schematic diagram of a pilot installation for conducting experiments on the purification of natural waters from nitrates (filtration modules with a doubled filtration flow path).

previous studies [26, 29] have shown that the so-called "Siroflock" process is convenient and applicable for further purification of natural waters from the drinking water supply system through local small-scale devices.

A development of a mechanical immedium filter for the so-called "Siroflock" process in dynamic conditions is presented, implemented through a mechanical filter with a specific constructive essence, where, in addition to the contact-coagulation interactions being governed by the positive charge of the magnetite particles and

Type of anions	Sample numbering with different water compositions			
	I	II	III	IV
HCO_3^-, mg/dm^3	140,3 ÷ 158,6	67,1 ÷ 91,5	MDL*	152,5 ÷ 164,7
SO_4^{2-}, mg/dm^3	180,0 ÷ 192,1	80,1 ÷ 95,6	264,2 ÷ 301,7	MDL
Cl^-, mg/dm^3	14,4 ÷ 18,0	14,4 ÷ 18,0	14,4 ÷ 18,0	14,4 ÷ 18,0
NO_3^-, mg/dm^3	56,8 ÷ 68,9	59,6 ÷ 65,3	58,6 ÷ 73,9	55,5 ÷ 68,1

**MDL – Minimum Detectable Limit.*

Table 4.
Anion composition of the studied model water solutions.

the dosed coagulant (chitosan), they are also influenced by the applied surrounding magnetic field (**Figure 7**). The reason is that the wider spectrum of the grain size of the filtration material, when stratified, creates a suitable inter-contact porosity of the bed, allowing the so-called "volume adhesion" filtration mechanism to occur, which results in a greater capacity of the bed compared to the conventional "surface" filtration mechanism. This also allows for higher linear filtration speeds and therefore a higher flow rate of the produced filtrate.

For comparison in both qualitative and quantitative aspects, laboratory experiments for coagulation treatment of water were conducted under volume coagulation conditions (through the so-called "Jar test") and under dynamic filtration conditions in a non-conventional filter with a conical body and surrounding magnetic field [15].

The body of the filtration device has the shape of an inverted cone. This design allows the processed water flow to enter from the side with the smaller diameter, while the filtrate (product) exits from the wider part. In this way, the device operates with permissible linear filtration speeds, higher flow rates, and lower hydraulic resistances. The construction includes a surrounding electromagnet (solenoid) that generates a solenoidal magnetic field, which envelops the top layer of the filtering ferromagnetic layer with the finest grain composition. This imparts magnetic properties to the magnetite particles, effectively ensuring better compaction of the filtration bed and thus a higher-quality filtrate at relatively high flow rates.

The advantage of this immedium filter is its ability to filter water at higher flow rates compared to conventional cylindrical immedium filters. This is possible because, as the water passes through the bed from the narrower part to the wider section of the device, its linear speed progressively decreases as it flows through the progressively expanding cross-section. Since the linear speed, and thus the force that tends to fluidize the upper layer of the filtering magnetite layer, progressively decreases along the direction of the outflowing stream, and due to the presence of the solenoidal field, the filtering layer remains compacted.

During the operation, the electrical power is supplied, and consequently, the surrounding electromagnetic field is present. Some of the coarse and colloidally dispersed impurities in the water are initially trapped in the first filtering layer along the path, which does not have ferromagnetic properties. The retention occurs based on adhesive, cohesive, and electrostatic interactions, with only the smaller colloidal impurities passing through the layer. These impurities are retained to a certain extent in the upper ferromagnetic layer, where the electromagnetic field facilitates magnetic coagulation. After a certain period of operation, due to the accumulation of retained impurities and the narrowing of the inter-contact spaces in the bed, the linear velocity of the water flow becomes greater than the forces holding the impurities onto the filtering grains. As a result, the filtration "front" shifts to the next filtering layer along the flow path. Simultaneously, the retained impurities in the respective lower filtration layer cause the fine colloidal impurities that were not trapped to grow through the so-called magnetic flocculation process and be retained in the final ferromagnetic magneto-structured layers of the filtration bed. This occurs due to the magnetic orientation of the impurities and their agglomeration through adhesion and cohesion processes.

After filtering a certain amount of treated water, due to the progressive movement of the main filtration "front" along the direction of the outflowing filtrate, the front reaches the final magnetically structured layer at the end of the flow path. Once the capacity of this layer is exhausted, a breakthrough greater than the initial (background) breakthrough established at the beginning of the filtration cycle appears in the filtrate flow. The system then enters the so-called recovery period of the filter, that is, it is introduced into a loosening process through intensive fluidization of the filtering material. This is achieved by removing the solenoidal magnetic field through interrupting the power supply. Part of the water in the filter body is drained, so that the bed remains submerged (with 30–40 cm of liquid above it). Air is supplied at an appropriate pressure, and the filtering layer is bubbled for about 5–6 minutes. The air flow is then stopped, and a flow of treated water is introduced from the bottom upward. In the absence of the magnetic field, the filtering material is fluidized for a certain period of time until the turbidity of the loosening flow equals that of the incoming flow. After completing the recovery loosening period, the flow is stopped for a few minutes. The filtering material is then gravitationally stratified based on particle size, the solenoidal magnetic field is reintroduced, and the filter returns to the working filtration period.

The results of comparative experimental studies for treating the same water using volume coagulation and lime decarbonation ("Jar test" via a laboratory apparatus) and using the above-mentioned mechanical filter with a conical body filled with magnetite sand are presented in **Figures 8–11**. The changes in the treated water quality are compared based on the following parameters: permanganate index (P.O.); humic acid concentration; turbidity; total hardness; and bicarbonate concentration (Am.o).

The reduction levels of the baseline parameters depicted in the figures range from 45 to 50%. **Figure 9**, which illustrates the variation of these same parameters during the operational period of the contact-coagulation filtration apparatus, demonstrates similar trends. With a suitable grain-size composition of the filtration medium, the

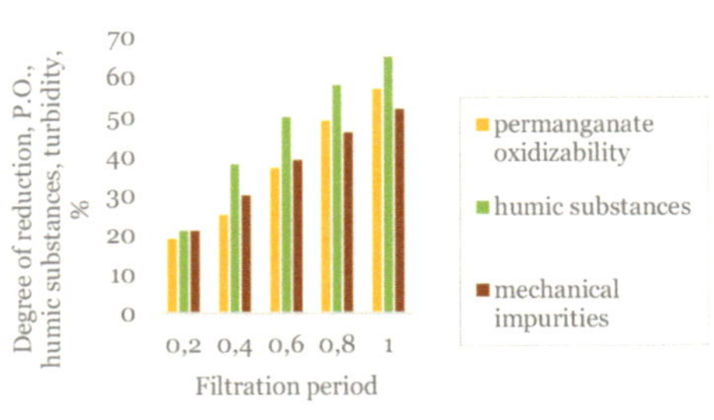

Figure 8.
Purification efficiency (%) evaluated based on: Permanganate index (P.O.), humic substances, and suspended solids over the course of the operational period, as determined by the "Jar test" using the aforementioned optimal doses of coagulant and pH correction reagent or lime decarbonation.

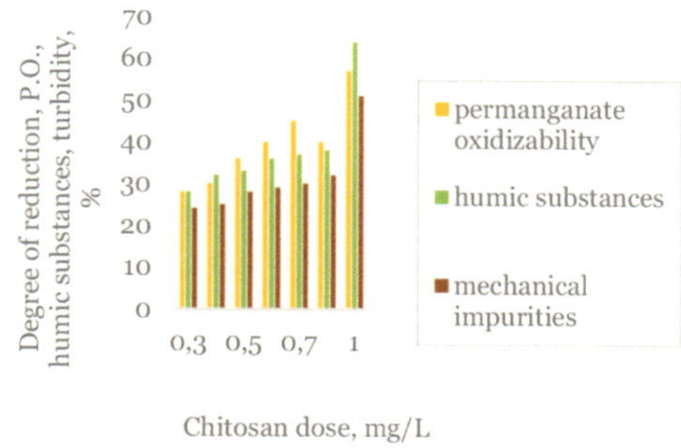

Figure 9.
Variation in purification efficiency (%) during coagulation "Jar test," evaluated based on: Permanganate oxidizability (P.O.), humic substances, and suspended solids, as a function of the coagulant dose (chitosan).

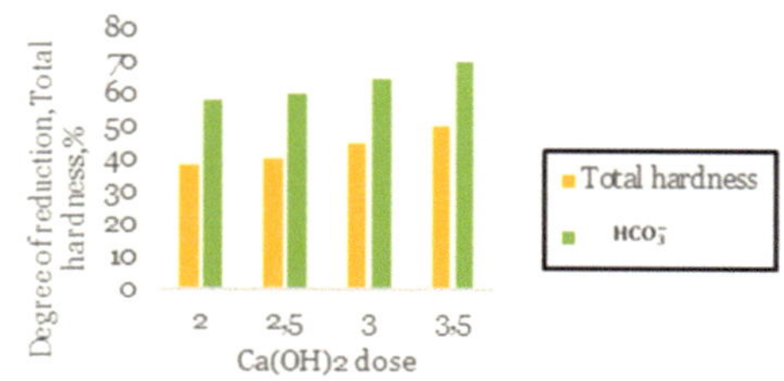

Figure 10.
Purification efficiency (%) during the "Jar test" for lime decarbonization, evaluated based on total hardness, humic substances, and suspended solids, as a function of the reagent dose for pH adjustment.

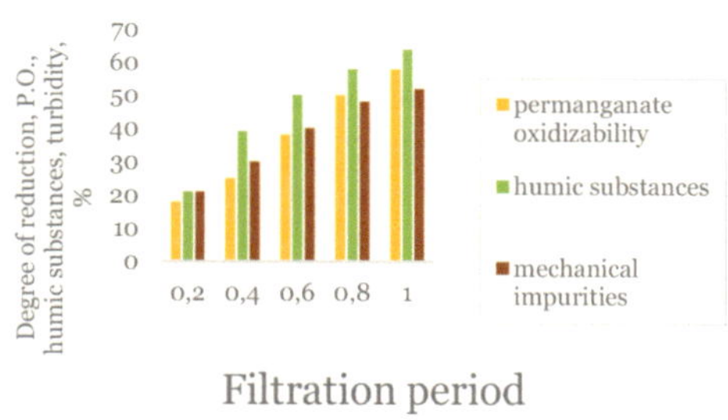

Figure 11.
Variation in purification efficiency (%) during the contact-coagulation process, based on permanganate oxidizability (P.O.), humic substances, and suspended solids, over the operational period of a mechanical filter filled with black magnetite sand (the "Siroflock process").

so-called adhesion-volume contact-coagulation filtration mechanism effectively removes colloidal and coarse-dispersed impurities from the water. The generated circumferential magnetic field prevents fluidization of the final sand-magnetite layers along the flow path, maintaining appropriate interstitial pore spaces. The pore dimensions, combined with magnetic flocculation processes, determine the quality of the treated water (filtrate).

It is concluded that this technological approach is more suitable and practical for small-scale water treatment facilities designed for localized additional purification of natural water for domestic use. The reason lies in the use of a compact filtration module that does not require reagent-based treatment. The presumed volumetric-adhesion contact-coagulation mechanism underlying the implemented "Siroflock" filtration process under dynamic filtration conditions is based on the attraction and adhesion/cohesion retention of colloidal humic substances on the magnetite sand grains due to their opposite electrostatic charges (natural humic organic matter carries a negative charge, while magnetite particles carry a positive charge). This adhesion/cohesion retention occurs throughout the entire volume of the magnetite filtration filling when an appropriately selected grain size composition of the filling is used.

6.1 Alternative option 1: Single-body three-chamber cylindrical filtration unit for local treatment of natural water for drinking purposes

The outer chamber is filled with quartz sand with a grain size of 0.5–1.5 mm and a layer height of H = 1000 mm. The first central chamber is filled with a mixture of weakly acidic cation exchange resin *Vofatite CA-20* in hydrogen ion form and activated carbon. The ratio of these two components is 1:1 but can vary depending on the quality of the natural water treated in the specific region and the target requirements for the treated water product delivered to consumers. The second chamber, located in the core center, is filled with activated carbon impregnated with elemental silver within its porous structure.

The construction allows the treated water flow to undergo mechanical filtration as it passes through the compressed sand layer and drainage nozzles. Via internal communication pipes, the flow continues downward through the compressed layers of the subsequent two chambers. Along its path, the water undergoes mechanical filtration and interacts via sorption with the activated carbon and weakly acidic cation exchange resin (enabling sorption and partial softening), and in the final central chamber, it undergoes additional purification and sterilization due to contact with activated carbon impregnated with elemental silver. The purified water is then discharged as a final product for the consumer (**Figure 12**).

During this treatment process, as the water passes through the described layers in the three consecutive chambers, the raw natural water undergoes significant improvements: its turbidity and permanganate oxidizability (representing coarse and colloidal organic impurities) are reduced; it is partially softened and decarbonized; and harmful impurities such as total iron (Fe total Mn^{2+}, halides) and others are reduced through sorption. Simultaneously, the treated water gains bactericidal properties.

The obtained results are presented by monitoring the changes in the concentrations of the aforementioned impurities, as shown in **Figure 13**.

It is evident from the data in **Figure 13** that the treated water reduces the content of certain impurities as follows: turbidity (83%), permanganate oxidizability (61%), total hardness (88%), manganese (92%), total iron (87%), and specific electrical conductivity (62%). The measured parameters meet the requirements specified in BDS 2823–83 and Ordinance No. 9/2001.

6.2 Alternative option 2: Single-body cylindrical three-chamber filtration system for simultaneous mechanical filtration and deep softening of water

The surrounding chamber is filled with quartz sand with a grain size distribution of 0.5–1.5 mm and a layer height of H = 1000 mm. The first and second central chambers along the water flow contain strong acid cation exchange resin Dowex HCR-S in sodium ion form, each with a layer height of H = 1000 mm. The treated water flow first passes through a sand layer for mechanical filtration and clarification. Then it

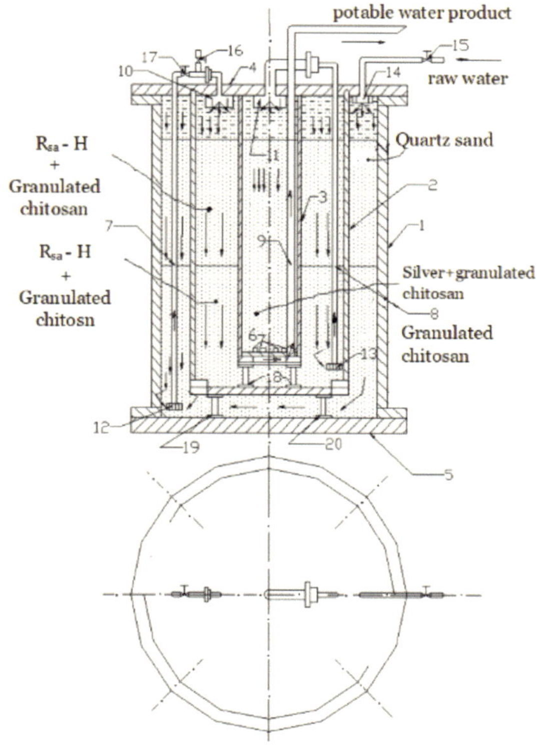

Figure 12.
Schematic diagram of a compact cylindrical single-body, three-chamber filter (for drinking water): 1 – outer casing; 2 – first inner cylinder; 3 – second inner cylinder; 4 – top cover; 5 – bottom cover; 6 – nozzles; 7, 8, 9 – internal connecting pipes; 10, 11, 12, 13, 14 – nozzles; 15, 16, 17 – valves; 18, 19, 20 – supporting stands [15].

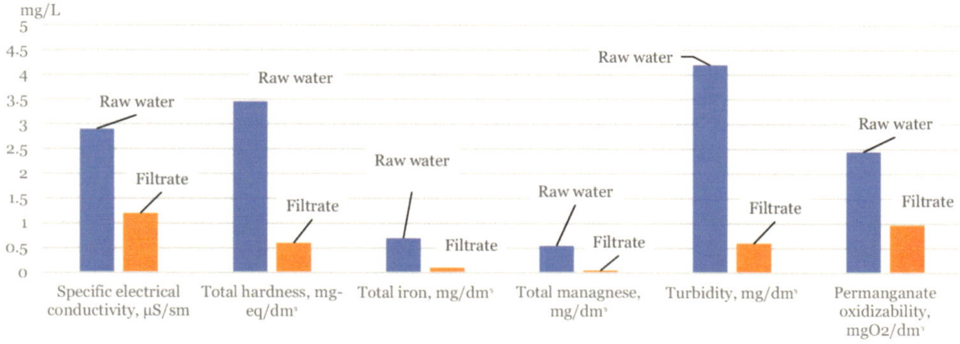

Figure 13.
Degree of reduction of various parameters of the treated water through a filtration experiment for local treatment of natural water with a pilot single-body, three-chamber filter.

flows through drainage nozzles and connecting pipes before being sequentially filtered top-down through the compressed Dowex HCR-S layers. During this two-stage cation exchange process, Ca^{2+} and Mg^{2+} ions are replaced with Na^{+} ions, resulting in deeply softened water. The treated water is then directed through a connecting pipe to the consumer (**Figure 14**).

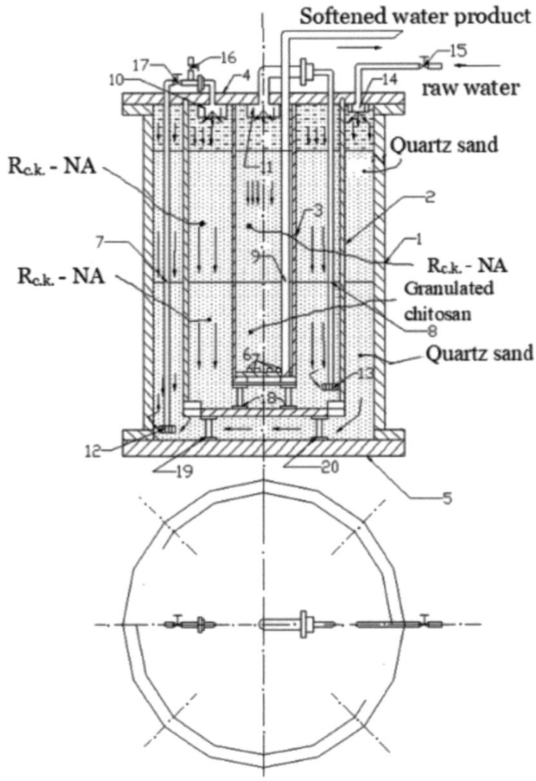

Figure 14.
Schematic diagram of the design of a compact cylindrical system.

Once breakthrough of the hardness cations above the pre-set permissible concentration occurs, the filtration system is switched to a regeneration mode to restore its technological functionality.

The different filtration media used in the two distinct chambers (sand and strong acid cation exchange resin) require a specific procedural approach to restore their working functionality. The following operational procedures are carried out:

1. Simultaneous loosening of the cationic and sand layers.

2. Reagent-regenerative process to restore the ion-exchange capacity (IEC) of the cationic media: When the quality of the effluent water at the end of the second (rapid) final backwash meets the required consumer standards (residual hardness), a new operational period begins. From this point, the filter exits the regeneration-restoration phase and enters the functional period, and the treated water is ready for consumption [28].

The results obtained from the experiments are presented in **Table 5**. The results shown in **Table 5** indicate that the three-chamber single-body filtration system, shown in **Figure 14**, can carry out two consecutive technological processes: mechanical filtration and ion-exchange softening. The first process clarifies the water to an acceptable residual turbidity, ensuring that the subsequent two filtration layers of cation exchange resin are protected from blocking their functional ion-exchange

Relative volume	Debit dm^3/h	Total hardness, $mg{-}eq/dm^3$		Turbidity, mg/dm^3		P.O., mgO_2/dm^3		Chlorides mg/dm^3	
		Raw water	Water product	Raw water	Water product	Raw water	Water product	Raw water	Water product
35	360	3,5	MDL*	4,4	1,0	2,3	0,6	10	10
70	360	3,4	MDL	4,2	0,9	2,1	0,7	12	12
105	360	3,6	MDL	4,5	0,9	2,0	0,5	10	10
140	360	3,5	MDL	4,4	0,9	2,4	0,5	9	9
175	360	3,5	MDL	4,3	1,0	2,1	0,4	10	10
210	360	3,4	MDL	4,1	0,9	2,3	0,5	12	12
245	360	3,5	MDL	4,3	0,8	2,2	0,4	11	11
280	360	3,4	MDL	4,5	0,8	2,3	0,4	9	9
315	360	3,3	MDL	4,2	0,8	2,5	0,3	10	10
350	360	3,5	MDL	4,3	0,8	2,1	0,4	12	12
385	360	3,2	MDL	4,1	0,8	2,4	0,3	11	11
420	360	3,5	MDL	4,5	0,8	2,1	0,4	10	10
455	360	3,6	0,01	4,2	0,8	2,2	0,4	10	10
490	360	3,3	0,02	4,3	0,8	2,3	0,3	9	9
525	360	3,5	0,03	4,1	0,8	2,1	0,3	10	10
560	360	3,4	0,10	4,4	0,8	2,5	0,4	11	11
595	360	3,5	0,25	4,3	0,8	2,0	0,3	10	10

**MDL. – Minimum Detectable Limit (MDL).*

Table 5.
Results regarding the quality of the product water during the treatment of raw natural water in alternative experiment 2; Vcat = 14 dm^3.

properties. Up to the treatment of about 420 relative volumes of water, the total hardness of the treated water is below the detectable minimum (MDL) with the Eriochrome Black-T indicator. Upon reaching the maximum permissible concentration (MPC) set by the consumer, the filter enters a regeneration period to restore its ion-exchange functionality. It is recommended to perform a loosening process for both the cation exchange resin and the quartz sand layer before this. The system's design is such that this process can be carried out sequentially by the same loosening flow.

In conclusion, regarding the reagent-regeneration restoration processes, it should be noted that such processes are not provided or carried out for the experiments in Alternative Option 1. In this case, the sorption-cationic mixture is replaced after the treatment of a certain volume of water. The size of this volume between two changes depends on the characteristics of the raw water being treated, that is, the degree of contamination, the frequency of accidental emergency infiltrations in the drinking water, repairs to the pipeline system, the region of the drinking water source, and so on. If there is a need for cascading (sequential) implementation of the same technological water treatment option (i.e., multiple treatment cycles for drinking purposes or repeated softening), the design developed by us allows for the connection of two or more filtration modules.

The aim of this type of fillings as mixed compositions is to remove sorption-toxic, harmful pollutants and to achieve a reduction in the content of hardness ions in the water to a certain extent. This results in clearer water with improved organoleptic qualities, a certain degree of softening, and ensured bacterial purity. The technological relationships and behavior in the process of usability have been studied using pilot small-scale water treatment installations. The obtained results are reflected in **Table 5**.

It is well known that the most commonly used sorption-filtration fillers include granular sorbents based on activated carbon, zeolites, ion-exchange resins, and polymeric or biopolymeric sorbents. The granules typically have a grain size characteristic of 1 ÷ 3 mm and are obtained by granulating a powdered starting product (chitosan, charcoal, zeolite, etc.) in the presence of a binding additive. In all cases, granular sorbents with so-called bidisperse porosity are formed [15].

The intercontact volumes between individual micro-particles with non-physical microporosity actually generate a secondary porosity (macroreticulation) with true physical porosity, referred to by the author as polyreticular.

The experimental results show that the mixture of sorbent (chitosan, activated carbon) and weakly acidic cationite in hydrogen ion form has a high purification capacity. It can treat approximately 1070 relative volumes of model water with iron and manganese concentrations 3–4 times above the maximum permissible limit (MPC) under BDS 2823–83. The results of the analyzes regarding the microbial count (M.C., number of bacteria/cm^3) show a minimal number of saprophytic bacteria in 1 ml of water. This is evidently due to the filtration contact of the treated water with the filler made of activated carbon implanted with elementary silver and/or chitosan.

Within the volume of the respective bodies, vertical pipes are placed to facilitate communication of the treated water and reagent-regeneration flows by allowing the passage of the reagent solution. The device can perform various alternative functions depending on the technological fillings in its chambers and their specific reagent treatment. **Figure 14** illustrates the fillings for the deionization functionality of the device.

6.3 Design-technological schemes for manufacturing small-scale water treatment devices

The third application alternative for the small-scale devices developed by us is the capability to deionize water through processes of ion exchange and final sorption. The resulting water (deionized water) must meet the requirements outlined in BDS ISO 3696 "Water for Laboratory Analysis," 1993; DIN 38404, 1952; and ASTM Designation: D 1193–99, 1999. The basic principle of the designed single-body three-chamber filter (as a deionizer) is shown in **Figures 15** and **16**. The filter consists of three nested cylindrical bodies (1, 2, and 3), forming three separate vertical filtration chambers, enclosed with upper and lower caps. The chambers are filled with filtration materials suitable for the technological purpose, forming compressed layers with a height of H = 1000 mm.

The design and technological solutions for developing treatment facilities for local pollutants from household, laboratory water flows, and wastewater from small-scale productions use two distinct chemical phenomena—ion exchange and adsorption. These processes are applied sequentially, starting with ion exchange capacity and then using the biopolymer chitosan as an adsorption material to facilitate adsorption processes and chemical transformations at the interface with

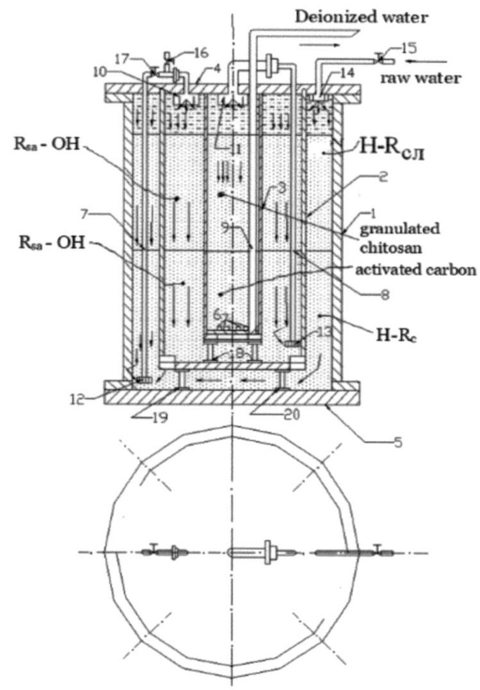

Figure 15.
Schematic design of a small-scale single-body three-chamber filter device (deionizer): 1 – outer body; 2 – first inner cylinder; 3 – second inner cylinder; 4 – upper cap; 5 – lower cap; 6 – nozzles; 7, 8, 9 – internal communication pipes; 10, 11, 12, 13, 14 – nozzles; 15, 16, 17 – valves; 18, 19, 20 – support stands [15].

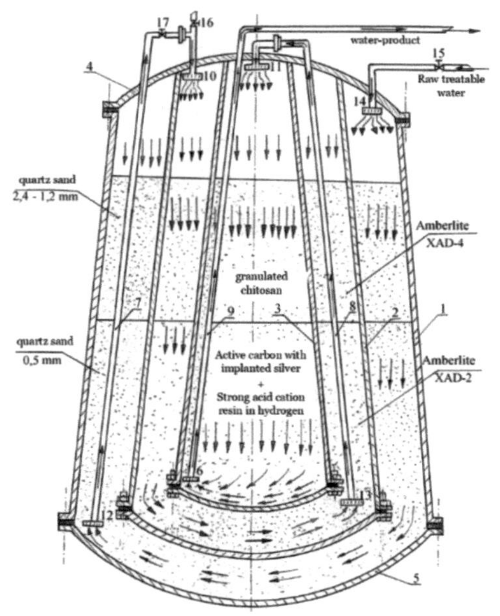

Figure 16.
Schematic design for a multi-chamber, multiprocess water treatment medium-scale or small-scale device with a conical body (alternative variant 1 – for additional purification of drinking water): 1 – outer body; 2 – intermediate body; 3 – inner body; 4 – upper cap; 5 – lower cap; 6, 10, 11, 12, 13, 14 – nozzles; 7, 8, 9 – communication pipes; 15, 16, 17 – valves [15].

the adsorbent. In this way, the treated water undergoes additional purification from organic and colored impurities. The incorporation of silver into the chitosan structure enhances the process by providing bactericidal protection for the treated water intended for household use.

All experimental data were processed using MATLAB statistical analysis meta-routine.

7. Conclusions

Compact multi-chamber and single-chamber deionization devices have been developed. At the end of their technological scheme, a sorption filler made of the biopolymer chitosan or a synthetic polymer sorbent is included. Their technological behavior has been studied in terms of the specific electrical conductivity of the product (via built-in flow-through conductometers) and the permissible concentrations of coarse-dispersed, colloidal-dispersed, molecular-dispersed, and other non-ionically dispersed impurities.

The developed constructions can have either cylindrical or conical shapes. Research indicates that the conical shape is preferable due to its advantages in process kinetics. Depending on the type of filtration material used, the device can perform various water purification functions:

- Mechanical filtration (three-stage, if designed solely as a mechanical filter);
- Ion-exchange softening, decarbonization (Na-R/R-Cl ionization), sorption with chitosan;
- Mechanical filtration, partial ion-exchange softening, and sorption with chitosan;
- Ion-exchange desalination, where the design allows for the inclusion of an additional filtration unit (in the case of a single-chamber, dual-chamber filter). Depending on the filler loaded into this additional unit, it can serve as a mixed-action filter for ion-exchange final water treatment, aiming to produce highly desalinated water with a specific electrical conductivity of less than 1 ms/cm according to BDS ISO 3696.

It is also possible for this additional filtration unit to function as a sorption treatment module, if it is filled with chitosan, activated carbon, or another polymer sorbent.

Attempts have been made to technologically design, construct, and fabricate prototype models of compact filtration-ion exchange deionization devices (consisting of multiple modules) for the production of deionized water intended for laboratory purposes or small-scale operations requiring such consumption. Depending on the required level of deionization and the quality parameters (e.g., salinity, turbidity) of the raw feedwater, installations with two or more filtration modules can be constructed. This is directly related to and significantly influences the technical and economic considerations when investing in the production of compact water treatment installations for deionized water.

Multihull deionizers facilitate the relatively straightforward implementation of all functional operation phases (working, loosening, reagent regeneration, slow rinsing, and rapid final rinsing). The reagent-regeneration filtration (dynamic) contact between the ion exchange media and the reagent solutions of appropriate concentration and volume is achieved through gravitational interaction, eliminating the mandatory need for a technician-operator's presence. These designs do not require pumps or other energy-consuming technical components. The necessary pressure for the treated feedwater flow is provided by the potable water supply system. A flow sensor for a digital conductivity meter is integrated at the outlet to monitor the quality of the produced water.

The proposed new design solutions for compact local water treatment, the study of material properties, and the development of new technological approaches for producing filtration, sorption, ion-exchange, and other types of media for small-scale water purification systems will increasingly contribute positively to their practical implementation in daily life. They will further demonstrate their applicability and ecological importance on a global scale. This, in turn, will help mitigate the serious environmental impacts on the quality of water in reservoirs used for drinking water supply. Combined with the availability of compact, multi-step, multiprocess filtration-sorption systems installed in homes, hospitals, kindergartens, resort complexes, and other facilities, these advancements will ensure access to healthy drinking water with good taste and quality.

To improve technological processes and schemes for local treatment of contaminated water using small-scale facilities, future scientific research could expand the scope of chitin and chitosan applications in two directions.

In the preliminary technological treatments of finely and coarsely dispersed impurities in water, studies incorporating chitin could contribute to enhancing the filtration process. One potential research area is exploring the ratio of different chitin fractions and quartz sand used in mechanical filtration. Knowing the broad spectrum of chitosan's properties, its potential applications could also be further expanded.

The protonation of various functional groups in the chitosan molecule at pH values in both acidic and alkaline ranges could expand its sorption capacity. This research area could contribute to designing and developing new multi-chamber structural solutions with an adaptive nature, depending on the composition of contaminants in the treated water. In this direction, conditions will be created for treating water with a diverse polyionic composition through chemical transformations. As a result, economic costs and the time required for the chemical treatment of water flows will be reduced.

Future research directions are related to investigating the long-term performance of chitosan under semi-industrial conditions when tested with a wider range of wastewater compositions. Extending the scope of research will contribute to improving wastewater treatment technologies and maintaining environmental sustainability.

Acknowledgements

I dedicate this chapter of the monograph to my husband and teacher Assoc. Prof. Eng. Angel Zvezdov, PhD.

Author details

Dilyana Zvezdova
Department of Pre-clinical and Therapeutic Studies, Faculty of Public Health and Health Care, Prof. Dr. Assen Zlatarov University, Burgas, Bulgaria

*Address all correspondence to: zvezdova@abv.bg

References

[1] Muna AM, Sreekrishnan TR. Aquatic toxicity from pulp and paper mill effluents: A review. Advances in Environmental Research. 2001;**5**(2):175-196. DOI: 10.1016/S1093-0191(00)00055-1

[2] Banat I, Nigam P, Singh D, Marchant R. Microbial decolourization of textile-dye-containing effluents: A review. Bioresource Technology. 1996;**58**(3):217-227. DOI: 10.1016/S0960-8524(96)00113-7

[3] Giles CH, Mc Kay RB. The lightfastness of dyes: A review. Textile Research Journal. 1973;**33**:528-577. DOI: 10.1177/004051756303300707

[4] Pokhrel D, Viraraghavan T. Treatment of pulp and paper mill wastewater—A review. Science of the Total Environment. 2004;**333**:37-58

[5] Robinson T, McMullan G, Marchant R, Nigam P. Remediation of dyes in textile effluent: A critical review on current treatment technologies with a proposed alternative. Bio/Technology. 2001;**77**:274-255. DOI: 10.1016/s0960-8524(00)00080-8

[6] Thompson G, Swain J, Kay M, Forster CF. The treatment of pulp and paper mill effluent: A review. Bioresource Technology. 2001;**77**:275-286. DOI: 10.1016/S0960-8524(00)00060-2

[7] Crini G. Non-conventional low-cost adsorbents for dye removal: A review. Bioresource Technology. 2006;**97**:1061-1085. DOI: 10.1016/j.biortech.2005.05.001

[8] Synowiecki J, Al-Khateeb NA. Production, properties, and some new applications of chitin and its derivatives. Critical Reviews in Food Science and Nutrition. 2003;**43**:145-171. DOI: 10.1080/10408690390826473

[9] Rinaudo M. Chitin and chitosan: Properties and applications. Progress in Polymer Science. 2006;**31**:603-632. DOI: 10.1016/j.progpolymsci.2006.06.001

[10] Struszczyk MH. Chitin and chitosan. Part II. Applications of chitosan. Polimery. 2002;**47**(6):396-403

[11] Zvezdova D. Application of biopolymer and synthetic polymer products. In: Atanasov A, Dimitrov G, editors. Burgas: Librascorp House; 2020. 420 p

[12] Zvezdova D. Production and characteristics of biopolymer and synthetic polymer products. In: Atanasov A, Dimitrov G, editors. Burgas: Librascorp House; 2021. 343 p

[13] Cheng Y, Wong R, Soo YOY, Wong WS, Lee CK, Ng MHL, et al. Use of convalescent plasma therapy in SARS patients in Hong Kong. European Journal of Clinical Microbiology & Infectious Diseases. 2005;**24**:44-46. DOI: 10.1007/s10096-004-1271-9

[14] Varma A, Deshpande SV, Kennedy JF. Metal complexation by chitosan and its derivatives: A review. Carbohydrate Polymers. 2004;**55**:77-93. DOI: 10.1016/j.carbpol.2003.08.005

[15] Zvezdova D, Zvezdov A. Organic biopolymer sorbent materials: Structural characteristics and their significance in various engineering technologies. In: Pavlova S, editor. Burgas: Librascorp House; 2014. 284 p

[16] No HK, Meyers SP. Application of chitosan for treatment of

wastewaters. In: Ware GW, editor. Reviews of Environmental Contamination and Toxicology. Berlin: Springer; 2000. pp. 1-27. DOI: 10.1007/978-1-4757-6429-1_1

[17] Ravi Kumar MNV. A review of chitin and chitosan applications. Reactive and Functional Polymers. 2000;**46**:1-27. DOI: 10.1016/S1381-5148(00)00038-9

[18] Berger J, Reist M, Mayer JM, Felt O, Peppas NA, Gurny R. Structure and interactions in covalently and ionically crosslinked chitosan hydrogels for biomedical applications. European Journal of Pharmaceutics and Biopharmaceutics. 2004;**57**:19-34. DOI: 10.1016/s0939-6411(03)00161-9

[19] Crini G, Martel B, Torri G. Adsorption of C. I. basic blue 9 on chitosan-based materials. International Journal of Environment and Pollution. 2008;**33**:1-4. DOI: 10.1504/IJEP.2008.020809

[20] Cestari AR, Vieira EFS, dos Santos AGP, Mota JA, de Almeida VP. Adsorption of anionic dyes on chitosan beads. 1. The influence of the chemical structures of dyes and temperature on the adsorption kinetics. Journal of Colloid and Interface Science. 2004;**280**:380-386. DOI: 10.1016/j.jcis.2004.08.007

[21] Krajewska B. Membrane-based processes performed with use of chitin/chitosan materials. Separation and Purification Technology. 2005;**41**(3):305-312. DOI: 10.1016/j.seppur.2004.03.019

[22] Chiou MS, Li HY. Adsorption behavior of reactive dye in aqueous solution on chemical cross-linked chitosan beads. Chemosphere. 2003;**50**:1095-1105. DOI: 10.1016/S0045-6535(02)00636-7

[23] Hebeish A, El-Shafei A, Sharaf S, Zaghloul S. Development of improved nanosilver-based antibacterial textiles via synthesis of versatile chemically modified cotton fabrics. Carbohydrate Polymers. 2014;**113**:455-462. DOI: 10.1016/j.carbpol.2014.06.015

[24] Gordana H, Knez Ž, Leitgeb M. Different preparation methods and characterization of magnetic maghemite coated with chitosan. Journal of Nanoparticle Research. 2013;**15**:6. DOI: 10.1007/s11051-013-1751-x

[25] Dobrevski I, Zvezdov A. Therminology and classification of ion-exchange resins and polymeric sorbents. In: Proceeding of the 23th International Symposium on Selectctive Polymeric Sorbents. Prague; 1982. pp. 18-25

[26] No HK, Meyers SP. Application of chitosan for treatment of wastewaters. In: Ware GW, editor. Reviews of Environmental Contamination and Toxicology. Berlin: Springer; 2000. pp. 1-27. DOI: 10.1007/978-1-4757-6429-1_1

[27] Demirbas A. Biodiesel from vegetable oils via transesterification in supercritical methanol. Energy Conversion and Management. 2002;**43**:2349-2356. DOI: 10.1016/S0196-8904(01)00170-4

[28] Dobrevski I. Technology of water - Part I. In: Yovchev M, editor. Sofia: Technique House; 1991. 308 p

[29] Petkov KT, Zvezdov A, Pavlova S, Zvezdova D. Siroflock process as a pretreatment technological method for a small ion-exchange desalination water deionizer. Journal of Environmental Protection and Ecology. 2007;**8**(1):43-49